Infrared Absorbing Dyes

TOPICS IN APPLIED CHEMISTRY

Series Editors: **Alan R. Katritzky, FRS**
Kenan Professor of Chemistry
University of Florida, Gainesville, Florida

Gebran J. Sabongi
Laboratory Manager, Encapsulation Technology Center
3M, St. Paul, Minnesota

BIOCATALYSTS FOR INDUSTRY
Edited by Jonathan S. Dordick

CHEMICAL TRIGGERING
Reactions of Potential Utility in Industrial Processes
Gebran J. Sabongi

THE CHEMISTRY AND APPLICATION OF DYES
Edited by David R. Waring and Geoffrey Hallas

HIGH-TECHNOLOGY APPLICATIONS OF ORGANIC COLORANTS
Peter Gregory

INFRARED ABSORBING DYES
Edited by Masaru Matsuoka

STRUCTURAL ADHESIVES
Edited by S. R. Hartshorn

A Continuation Order Plan is available for this series. A continuation order will bring delivery of each new volume immediately upon publication. Volumes are billed only upon actual shipment. For further information please contact the publisher.

Infrared Absorbing Dyes

Edited by
Masaru Matsuoka
University of Osaka Prefecture
Sakai, Osaka, Japan

Plenum Press • New York and London

Library of Congress Cataloging-in-Publication Data

Infrared absorbing dyes / edited by Masaru Matsuoka.
    p.   cm. -- (Topics in applied chemistry)
   Includes bibliographical references and index.
   ISBN 0-306-43478-4
   1. Dye lasers--Materials.   I. Matsuoka, Masaru, 1942-
II. Series.
TA1690.I53 1990
621.36'2--dc20

                                               90-40052
                                                 CIP

ISBN 0-306-43478-4

© 1990 Plenum Press, New York
A Division of Plenum Publishing Corporation
233 Spring Street, New York, N.Y. 10013

Printed in the United States of America

Contributors

David Dolphin, Department of Chemistry, University of British Columbia, Vancouver, British Columbia, Canada V6T 1Y6

Atsushi Kakuta, Hitachi Research Laboratory, Hitachi Ltd., Hitachi-shi, Ibaraki-ken, 319-12, Japan

Fumio Matsui, Corporate Research and Development Laboratory, Pioneer Electronic Corporation, Tsurugashima-machi, Iruma-gun, Saitama 350-02, Japan

Masaru Matsuoka, Department of Applied Chemistry, College of Engineering, University of Osaka Prefecture, Sakai, Osaka 591, Japan

Yuji Mihara, Ashigara Research Laboratories, Fuji Photo Film Co., Ltd., Minami-Ashigara, Kanagawa 250-01, Japan

Yoshiharu Nagae, Hitachi Research Laboratory, Hitachi Ltd., Hitachi, Ibaraki 319-12, Japan

Kenryo Namba, Advanced Materials Research Department, R & D Center, TDK Corporation, Ichikawa, Chiba 272, Japan

Jun'etsu Seto, Research Center, SONY Corporation, Hodogaya-ku, Yokohama, Kanagawa 240, Japan

Ethan Sternberg, Department of Chemistry, University of British Columbia, Vancouver, British Columbia, Canada V6T 1Y6

Yoshiaki Suzuki, Late of Ashigara Research Laboratories, Fuji Photo Film Co., Ltd., Minami-Ashigara, Kanagawa 250-01, Japan

Tadaaki Tani, Ashigara Research Laboratories, Fuji Photo Film Co., Ltd., Minami-Ashigara, Kanagawa 250-01, Japan

Masayasu Ueno, Opto-Electronics Research Laboratory, NEC Corporation, Miyamae-ku, Kawasaki, Kanagawa 213, Japan

Tonao Yuasa, Opto-Electronics Research Laboratory, NEC Corporation, Miyamae-ku, Kawasaki, Kanagawa 213, Japan

Preface

New laser technology has developed a new dye chemistry! Development of the gallium–arsenic semiconductor laser (diode laser) that emits laser light at 780–830 nm has made possible development of new opto-electronic systems including laser optical recording systems, thermal writing display systems, laser printing systems, and so on. Medical applications of lasers in photodynamic therapy for the treatment of cancer were also developed. In such systems, the infrared absorbing dyes (IR dyes) are currently used as effective photoreceivers for diode lasers, and will become the key materials in high technology. At the present time the chemistry of IR dyes is the most important and interesting field in dye chemistry.

Laser light can be highly monochromatic, very well collimated, coherent, and, in some cases, extremely powerful. These characteristics make diode lasers a very cheap, convenient, and useful light source for a variety of applications in science and technology. For these purposes, however, IR dyes with special characteristics are required. To develop new IR dyes, it is most important to establish the correlation between the chemical structures of dyes and other characteristics of dyes, such as their absorption spectra.

Molecular design of IR dyes predicting the λ_{max} and ε_{max} values by molecular orbital (MO) calculations is now possible even by using a personal computer, and many types of new IR dyes have been demonstrated. Also, new opto-electronic systems using IR dyes as photoreceivers have been reported recently.

This book reviews the synthesis and characteristics of IR dyes and their applications in high technology. In Chapter 2, synthetic design of IR dyes by means of the PPP MO method is described. General methodology for the molecular design of IR dyes is demonstrated for each of the chromophoric systems. In the following six chapters, synthesis and characteristics of IR dyes from cyanine, quinone, phthalocyanine, metal complexes, and miscellaneous chromophores are described together with photochromic dyes. Chapter 9 deals with semiconductor lasers as a light source, and the remaining chapters deal with the applications of IR dyes in optical recording systems (Chapter 10), thermal writing displays (Chapter 11), laser printing systems (Chapter 12), laser filter systems (Chapter 13), infrared photography (Chapter 14), and medical applications (Chapter 15). In these chapters, the up-to-date results and technology are included and many figures are added for better understanding.

This book is intended mainly for color chemists, organic chemists, and material scientists. It is hoped that it will also be useful for postgraduate students in chemistry and material science.

It is a pleasure to gratefully acknowledge the contributors of each chapter. Without their hard work and kind cooperation, this book could not have been published in a timely manner. I also would like to record my special gratitude to the late Mr. Yoshiaki Suzuki who died shortly after the submission of his manuscript.

Finally, I would like to give my sincere thanks to Dr. John Griffiths of the University of Leeds, and the Series Editors, Drs. G. Sabongi and A. Kratitzki, who recommended me as an editor of this volume. I am also grateful to Miss Kazuko Shirai for her skilled assistance in preparing the manuscript.

Masaru Matsuoka

Osaka, Japan

Contents

Chapter 8.   Other Chromophores   89
Masaru Matsuoka

Part II.   Applications of Infrared Absorbing Dyes

Chapter 9.   Semiconductor Lasers   97
Masayasu Ueno and Tonao Yuasa

Chapter 10. Optical Recording Systems   117
Fumio Matsui

Chapter 11. Thermal Writing Displays   141
Yoshiharu Nagae

Chapter 15. Medical Applications

Ethan Sternberg and David Dolphin

Infrared Absorbing Dyes

1

Introduction

MASARU MATSUOKA

There is much current interest in the development of new infrared absorbing dyes for use as key materials in optoelectronics systems.[1] The development of the gallium–arsenic semiconductor laser, which emits laser light at 780–830 nm, enabled the development of new optoelectronic systems such as laser optical recording systems, thermal writing displays, and laser printing systems. Medical application systems in photodynamic therapy for the treatment of cancer were also recently developed with the use of lasers as the light sources. In such systems, the infrared absorbing dyes are used as effective photoreceivers for laser light.

Some cyanine infrared absorbing dyes have been known for some time to be useful in photochemistry, but their use was very restricted. The development of new types of infrared absorbing dyes has therefore been anticipated as a source of functional materials for high-technology applications. The chemistry and applications of infrared absorbing dyes will become one of the most interesting and important fields in the study of new dye chemistry. Infrared absorbing dyes are very new classes of dyes developed in the past decades.

In recent years, the focus of research on dye chemistry has largely changed from involvement in the traditional chemistry of dyes and pigments to investigation of special dyes for electro-optical applications. Over 130 years have passed since Perkin obtained the first commercial synthetic dye,

MASARU MATSUOKA • Department of Applied Chemistry, College of Engineering, University of Osaka Prefecture, Sakai, Osaka 591, Japan.

1

mauveine, in 1856. During this time, knowledge regarding the development of new dyes has accumulated and very many dyes have been synthesized.

Dyes have traditionally been used as coloring matters for polymer substrates such as textiles and plastics, whereas in high-technology applications, dyes are used as key materials which absorb light efficiently. Light sources used normally emit single-wavelength light, such as laser light, or narrow-wavelength light, such as that produced by a halogen lamp. Laser light can be highly monochromatic, very well collimated, and coherent, and, in some cases, laser sources are extremely powerful. These characteristics make the laser a very useful light source for a variety of applications in science and industry. For high-technology applications, dyes with special characteristics are required. To develop new dyes, it is most important to know the relationship between the chemical structures of the dyes and characteristics such as their absorption spectra.

In principle, dyes are extremely versatile materials and may be utilized in many ways. The exploitable properties of a dye chromophore have been summarized by Griffiths.[2] They are light absorption, light emission, light-induced polarization, photoelectrical properties, and chemical and photochemical reactivities. He pointed out that most of these properties are related to the ability of dyes to interact strongly with visible electromagnetic radiation, leading to such phenomena as color, fluorescence, and various photochemical and photoelectrical processes.

Infrared absorbing dyes can be applied in the following fields[3]: laser optical recording systems, laser printing systems, laser thermal writing displays, infrared photography, and medical or biological applications.

To develop new infrared absorbing dyes, it is essential to be able to correlate the characteristics of the dyes with their structures. A new methodology for the development of new special dyes has been reported recently.[4] It consists of two parts: the route for the development of new special dyes, and the four software schemes applied at each stage. There are many "needs" or demands in the development of new dyes as the key materials for electro-optical applications, for example. Some basic chromophores which have suitable physical properties are selected, and properties such as λ_{max}, ε, melting point, and so on are determined by using the dye database.[5] The database contains information on some 6,000 dyes, and each entry gives the chemical name, dye number, λ_{max} and ε values in several solvents, melting point, some other properties such as fluorescence, isomerism, sublimation, and toxicology and also end uses in some cases together with references to the original literature. The database is accessed by using a personal computer (NEC 9800, Toshiba J-3100, and IBM PC series), which makes it convenient to look at proposed or imagined dye structures. A useful data book containing information on some 2700 dyes has been recently published[6] and will prove to be a handy reference, contributing to the development of new dyes for nontextile use as well as serving as a useful guide to dyes used as traditional colorants.

After the selection of dye chromophores by using the database, the molecular design of dyes is based on the molecular orbital (MO) calculation method. If necessary, the molecular mechanics (MM) method to determine the most favorable configuration of the molecule and the molecular dynamics (MD) method to evaluate the molecular interactions in aggregates or crystals can be applied. Great advances in knowledge about the color-structure relationships of dye chromophores are attributable to the development of the Pariser–Parr–Pople (PPP) MO theory.[7] Quantitative MO theory can be applied to the design of dye chromophores in terms of predicting the color and other color-related properties of dyes. The PPP MO calculation method analyzes chromophoric systems of dyes so that the substituent effect on the absorption spectra can be evaluated quantitatively. The molecular structure of any type of dye is now accessible via absorption spectra by applying the PPP MO method[8] (see Chapter 2, Section II). The proposed dye is then synthesized and its properties evaluated. The next stage is an appraisal of the dye's characteristics, and the results are fed back for reevaluation. The design is then modified accordingly, and the dye is synthesized again to obtain more desirable characteristics. This methodology can be used to develop new infrared absorbing dyes for a variety of applications.

Recently, the spectral charts of some 190 infrared absorbing dyes for diode lasers have been summarized and published as a data book.[9] Their absorption characteristics can be widely applied to the design of functional materials for optoelectronic devices.

This book consists of two parts. In Part I, syntheses and characteristics of infrared absorbing dyes are reviewed. Part II concerns the application of infrared absorbing dyes in high technology.

REFERENCES

1. M. Matsuoka, in: *Technology and Materials for Optical Recording Systems* (R&D No. 75), p. 176, CMC, Tokyo (1985) (in Japanese).
2. J. Griffiths, *Chemistry in Britain*, 997 (1986).
3. M. Matsuoka, *J. Soc. Dyers Colourists 105*, 167 (1989).
4. M. Matsuoka, in: Chemistry of Functional Dyes, eds. Z. Yoshida and T. Kitao, Tokyo, Japan p. 9 (1989).
5. M. Matsuoka, K. Shirai, and M. Morita, Abstracts of the 10th International Color Symposium, Trier, Federal Republic of Germany, P-18 (1988). *Chemical and Engineering News* p. 25, Dec. 18 (1989).

6. M. Okawara, T. Kitao, T. Hirajima, and M. Matsuoka, *Organic Colorants, A Handbook of Data of Selected Dyes for Electro-Optical Applications*, Kodansha-Elsevier, Tokyo (1988).
7. R. Pariser and R. G. Parr, *J. Chem. Phys. 21*, 466, 767 (1953); J. A. Pople, *Trans. Faraday Soc. 49*, 1375 (1953).
8. J. Fabian and H. Hartmann, *Light Absorption of Organic Colorants*, Springer-Verlag, Berlin (1980).
9. M. Matsuoka, *Absorption Spectra of Dyes for Diode Lasers*, Bunshin, Tokyo (1990).

I

Synthesis and Characteristics of Infrared Absorbing Dyes

2

Synthetic Design of Infrared Absorbing Dyes

MASARU MATSUOKA

I. INTRODUCTION

The color–structure relationship is the most important factor for the design of infrared absorbing dyes. Infrared absorbing dyes, which have been of interest recently in the development of functional materials for diode-laser technologies, comprise a very new category of dyes, and their synthetic design should be based on the new ideas and methodology discussed in Chapter 1.

The color–structure relationship of dyes was first rationalized in 1876 in terms of the chromogen theory, which established the basis of dye chemistry. The resonance theory established by Bury[1] in 1935 led to the development of the chemistry of π-electron and aromatic systems, which contributed greatly to the development of new synthetic dyes, and almost all of the dye chromophores commercialized today were developed by using the concepts of resonance theory. The resonance theory can be applied qualitatively to evaluate the chromophoric system, but the quantitative explanation of absorption spectra is required for the design of infrared absorbing dyes. The λ_{max} of a dye as a photoreceiver must be predicted correctly in order to apply dye materials for diode lasers, which emit single-wavelength laser light at 780-830 nm. Great advances in the quantita-

MASARU MATSUOKA • Department of Applied Chemistry, College of Engineering, University of Osaka Prefecture, Sakai, Osaka 591, Japan.

7

tive prediction of absorption spectra of dye chromophores are attributable to the development of the Pariser–Parr-Pople molecular orbital (PPP MO) theory in 1953.[2] The MO theory can be applied to the design of dye chromophores in terms of predicting color properties and other color-related properties of dyes. The PPP MO calculation method analyzes chromophoric systems of dyes so the substituent effect on the absorption spectra can be evaluated quantitatively. The molecular design of any type of dyes is now accessible via absorption spectra by application of the PPP MO method. The advanced MO calculation methods such as CNDO, MNDO, and *ab initio* are now becoming available for the prediction of the absorption spectra in much greater detail, but they are not simple and need longer calculation times.

In this chapter, the application of the PPP MO method to the design of infrared absorbing dyes is described.

II. PPP MO METHOD

The absorption spectra of any dye chromophore can be quantitatively calculated by the PPP MO method. It is similar to the simplest Hückel molecular orbital (HMO) method; both of them deal with only the π electrons of molecules, independently of the σ electrons. The PPP MO method differs from the HMO method in considering the electron repulsion effects. The self-consistent field method is applied in the PPP MO method: an approximate set of the linear combination of atomic orbital (LCAO) coefficients are first obtained by the HMO calculation method and then these are improved by iterative calculations evaluating the electron repulsion energy until no further improvements in the set of LCAO data are achieved. The resultant MOs are then said to be self-consistent. Transition energies can then be calculated from the orbital energies and the electron repulsion terms. At this stage, the calculated transition energies still do not reproduce the observed values. The configuration interaction (CI) treatment is applied to give better calculated values. The CI treatment mixes the various excited electronic configurations; in most cases, the five highest occupied MOs and the five lowest unoccupied MOs are considered for sets of configurations to give a set of the excited states.

The energy difference between the ground state and the first excited singlet state gives the transition energy for the first absorption band, which usually corresponds to a single-electron transition from the highest occupied molecular orbital (HOMO) to the lowest unoccupied molecular orbital (LUMO). Similarly, the second absorption band corresponds to the transition from the next highest occupied molecular orbital to the LUMO or that from the HOMO to the next lowest unoccupied molecular orbitals.

Each of the transition energies can be calculated similarly, and the calculated values reproduce the observed values quite well. The use of the PPP MO method is restricted to chromophores with coplanar structures. Steric effects can usually be considered for evaluating the nonplanar geometry of molecules.

Semiempirical values for the valence state ionization potentials, the electron affinities, or the one-center repulsion integrals for atoms are generally used in the PPP MO method, and these parameters have been generalized already.[3] These parameters can be applied to a variety of dye chromophores.

Some sets of PPP MO calculation programs that can be run on a personal computer are now available for design of dye chromophores.[4] They are set up automatically by including structure drawing and parameter setting, and then parts of desirable results can be printed. Several minutes are required for the calculation for a dye molecule of moderate size, for example, three minutes for phthalocyanine.

Advanced MO calculation methods, such as the MNDO or the CNDO method, which deal with the bonded σ-electrons together with the π-electrons of molecules are also available for predicting the absorption spectra of dye chromophores, but they need much longer calculation times and the results are not improved substantially. It can be said that the PPP MO method is satisfactory and convenient method for the design of dye chromophores on the basis of spectral properties.

The early applications of the PPP MO method for dye chromophores have been summarized by Griffiths[5] in *Colour and Constitution of Organic Molecules*, published in 1976, and practical applications of it for various dye chromophores have been discussed by Fabian and Hartmann[6] in *Light Absorption of Organic Colorants*, which appeared in 1980. Recently, Tokita et al.[4] published a book in Japanese entitled *Design of Functional Dyes by the PPP MO Method*, which gives an introductory treatment of PPP MO calculations. They summarized the parameters and dealt with practical examples for the design of some functional dyes. The practical applications of the PPP MO method in the design of some infrared absorbing dyes are described in the following sections.

III. SYNTHETIC DESIGN OF INFRARED ABSORBING DYES BY THE PPP MO METHOD

A. Intramolecular Charge-Transfer Chromophores

Intramolecular charge-transfer (CT) chromophores can be defined as chromophores in which movement of π-electron density from the donor to

Table I. PPP MO Calculation Results and Observed λ_{max} of Naphthoquinones[a]

No.	Substituent	Calculated[b]		Observed	
		ε_{HOMO} (eV)	ε_{LUMO} (eV)	λ_{max} [c] (nm)	$\Delta\lambda$ [d] (nm)
1		10.52	3.64	335	—
2	2,3-(CN)$_2$	10.79	4.48	365	30
3	5-NH$_2$-2,3-(CN)$_2$	9.45	4.33	585	220
4	5-NH$_2$-8-NHPh-2,3-(CN)$_2$	8.50	4.08	735	150
5	5-NH$_2$-8-NHPh-2-CN-3-NHBu	8.19	3.29	605	−130
6	5-NH$_2$-2-CN-3-NHBu	9.28	3.52	484	−121

[a] Ref. 8.
[b] Energy levels of HOMO and LUMO.
[c] λ_{max} values denote the position of the central peak of three peaks in the first absorption band.
[d] Difference in λ_{max} from the compound in the previous row.

the acceptor accompanies the first excitation. Examples are generally found in quinoid, azo, and indigo chromophores. A typical example is found in quinoid chromophores. The first absorption band corresponds to a single-electron transfer from the HOMO to the LUMO of the molecule, and the substituent effects can be evaluated in terms of the energy differences between the HOMO and the LUMO; that is, the energy levels of the HOMO and the LUMO are affected by the introduction of substituents.

The results of PPP MO calculations for naphthoquinones are summarized in Table I. It is well known that the first visible absorption band of 1,4-naphthoquinonoid dyes can be assigned to a benzenoid band of intramolecular CT character.[7] A good linear correlation between the observed first excitation energy (ΔE_{max}) and the PPP MO calculated values (ΔE_1) for the first excitation was found[8] as shown in Figure 1 for a series of naphthoquinone dyes, and Eq. (1) was derived.

$$\Delta E_{max}(eV) = 1.06 \times \Delta E_1 - 0.382 \qquad (1)$$

A different linear plot with a similar slope was also obtained for another series of 1,4-anthraquinonoid dyes.[9] The MO results for the first excitation energy satisfactorily reproduce the trend in the observed values. A good

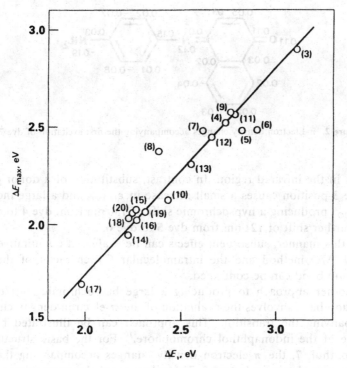

Figure 1. Relationship between the calculated excitation energy (ΔE_1) and the observed value (ΔE_{max}) for the benzenoid band of 1,4-naphthoquinone dyes.

linear correlation also exists between the first excitation energy and the singly excited configuration energy ($E_{LUMO-HOMO}$) accompanying the excitation from the HOMO to the LUMO. This shows that ΔE_1 depends markedly on the character of the HOMO and LUMO, and then the substituent effects can be discussed in terms of the ΔE_1 values and the energy levels of the HOMO and LUMO.

Typical substituent effects on ε_{HOMO} and ε_{LUMO} are shown in Table I. The following general observations have been made.[8] Substitution of the strong acceptor CN into the 2- and 3-positions of dye **1** causes a small decrease in ε_{HOMO} and a large decrease in ε_{LUMO}; that is, increasing the strength of an acceptor produces a bathochromic shift of 30 nm in the intramolecular CT transition. Substitution of a strong donor group (e.g., NH$_2$ or NHPh) into the 5- and/or 8-positions causes a large increase in ε_{HOMO} and a small increase in ε_{LUMO}; that is, increasing the strength of the donor group produces a large bathochromic shift of 220 nm from dye **2** to dye **3** and an additional shift of 150 nm from dye **3** to dye **4**, and dye **4**

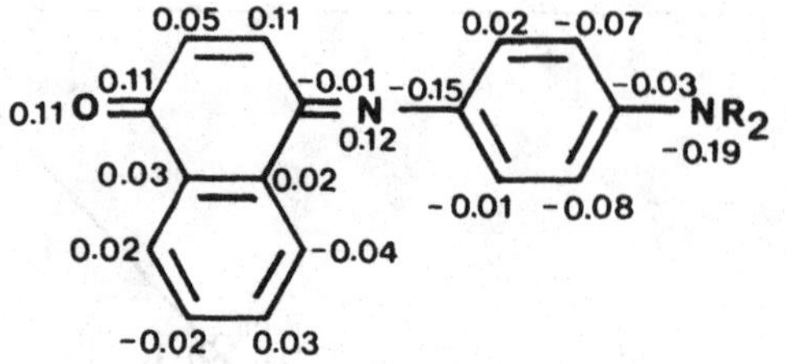

Figure 2. π-Electron density changes accompanying the first excitation of dye 7.

absorbs in the infrared region. In contrast, substitution of a donor group into the 3-position causes a small increase in ε_{HOMO} and a large increase in ε_{LUMO}, producing a hypsochromic shift of 130 nm from dye 4 to dye 5, and a further shift of 121 nm from dye 5 to dye 6.

In this manner, substituent effects can be evaluated quantitatively by the PPP MO method and the intramolecular CT character of the first absorption band can be confirmed.

Another approach to producing a large bathochromic shift of the absorption band involves the evaluation of the π-electron density changes accompanying the transition. This approach can be illustrated by the example of the indonaphthol chromophore.[10] For the basic structure of indonaphthol, 7, the π-electron density changes accompanying the first transition are shown in Figure 2. The values show the intramolecular CT

Table II. Substituent Effects on λ_{max} for the First Excitation of Dye 7[a]

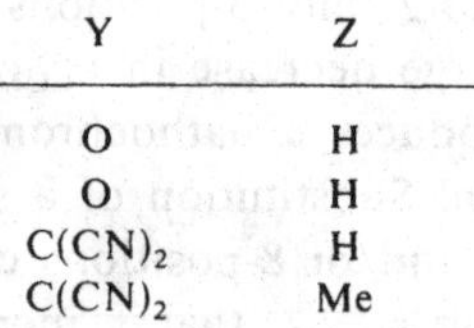

Dye	X	Y	Z	λ_{max} (nm)	$\Delta\lambda$[b] (nm)
7	H	O	H	583	—
8	CN	O	H	795	212
9	H	C(CN)$_2$	H	750	167
10	H	C(CN)$_2$	Me	762	179

[a] Refs. 10 and 11.
[b] Difference in λ_{max} from dye 7.

character of dye 7; that is, the aniline moiety acts as a donor and the naphthoquinoneimine moiety acts as an acceptor. From these results, substitution of acceptors at the 2- and/or 3-positions or substitution of a carbonyl group by much stronger acceptors such as dicyanomethide may be expected to produce large bathochromic shifts in λ_{max}.[11] On the other hand, substitution of donors at the 2'- and/or 5'-positions may also be expected to produce bathochromic shifts. In fact, dyes 8 and 9 absorb in the infrared region and exhibit bathochromic shifts of 212 and 167 nm, respectively, relative to dye 7 (Table II). Dye 10 absorbs at much longer wavelengths than dye 9. The bathochromic shifts for these dyes can be evaluated quantitatively by PPP MO calculations.

B. Intermolecular Charge-Transfer Chromophores

The first excitation energy ($\Delta E_{1(CT)}$) of the intermolecular CT chromophore can be correlated by Eq. (2); ΔE_1 is proportional to the differences between ε_{LUMO} of acceptor (A) molecules and ε_{HOMO} of donor (D) molecules.

$$\Delta E_{1(CT)} \propto \varepsilon_{LUMO}(A) - \varepsilon_{HOMO}(D) \qquad (2)$$

The frontier energy levels of a donor and an acceptor can be calculated independently by the PPP MO method. A linear relationship between ΔE_1 and the observed ΔE_{max} has been obtained in the cases of the carbazole-naphthoquinone intermolecular CT complex dyes.[12]

Equation (3) was deduced from the results:

$$\Delta E_{max}(eV) = 1.015 \times \Delta E_1 - 2.352 \qquad (3)$$

The calculation does not involve the configuration interaction of orbitals between donor and acceptor molecules, and the second excitation energy, ΔE_2, cannot be estimated.

The color-structure relationship of the intermolecular CT complex dyes can be established to be the same as in the cases of the intramolecular CT dyes by the PPP MO method. Infrared absorbing dyes can be designed by increasing the ε_{HOMO} of donor molecules and decreasing the ε_{LUMO} of acceptor molecules. In the cases of carbazole-naphthoquinone complexes, donor properties increase in the order NR > NH > Se > S > O in X, and acceptor properties are increased by the introduction of cyano and/or nitro groups into the naphthoquinone nucleus. It is also necessary to consider the morphological aspects of donor and acceptor molecules; both molecules have planar structures, and there are no steric interactions that can contribute to complex formation.

Donor	Acceptor
X = NR, NH, Se, S, O	X = Cl, CN; Y = H, NO$_2$

C. Cyanine Chromophores

The basic chromophoric systems of cyanine dyes have been rationalized in terms of the Dewar–Knott rule,[13,14] as described by Griffiths.[5] Klessinger[15] reevaluated the cyanine chromophores by the PPP MO method, and the results were summarized by Fabian and Hartmann.[6]

The chromophore for the symmetric cyanine dyes is a conjugated ionic type with odd alternant systems, and two extreme resonance structures are equivalent in energy and there is no C—C bond alternation in the ground and the excited states.

Dewar's rule can be applied to rationalise the substituent effect on the absorption spectra.[15] That is, a bathochromic shift is produced by the introduction of an electron-donating group at a starred position and/or an electron-withdrawing group at an unstarred position. On the other hand, a hypsochromic shift is produced by the introduction of an electron-with-drawing group at a starred position and/or an electron-donating group at an unstarred position.

Shifts of λ_{max} to longer wavelengths are caused by steric hindrance. When the twisting of a C—C bond occurs, a small bathochromic shift occurs if the number of atoms constituting the cyanine type chromophore is represented by $(4n + 1)$, and a large bathochromic shift occurs if it is $(4n - 1)$.

Yasui et al.[16] studied the substituent effects of trimethine cyanines on the basis of the PPP MO method and optimized the parameters for the PPP MO calculations. The π-electron density changes accompanying the first excitation of the trimethine cyanine dye (11) are shown in Figure 3. The π-electron densities decrease at the starred positions and increase at the unstarred positions as deduced by Dewar's rule. The dependences of the calculated λ_{max} of 11 on the IP values for the nitrogen atom and the heteroatom (X) are shown in Figure 4. The λ_{max} shows a large dependence on the IP value of the nitrogen atom but a small dependence on that of X.

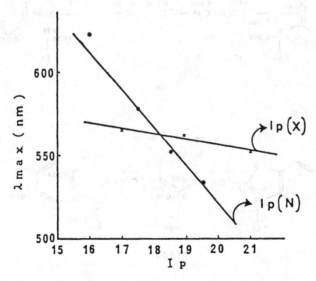

(11)

Figure 3. π-Electron density changes accompanying the first excitation of trimethine cyanine dye **11** (X = S, R = Et).

These results can be applied for the synthetic design of infrared absorbing dyes. It is generally known that the λ_{max} of cyanine dyes $(R+CH{=}CH+_n CH{=}R^+)$ exhibits a bathochromic shift of about 100 nm with an increase in n of 1 and shifts to the near-infrared region for $n > 3$. The λ_{max} value is also strongly affected by the nature of the aromatic moiety R. These results are summarized in detail by Griffiths.[5]

Figure 4. Dependences of the calculated λ_{max} of dye **11** (X = S, R = Et) on IP values.

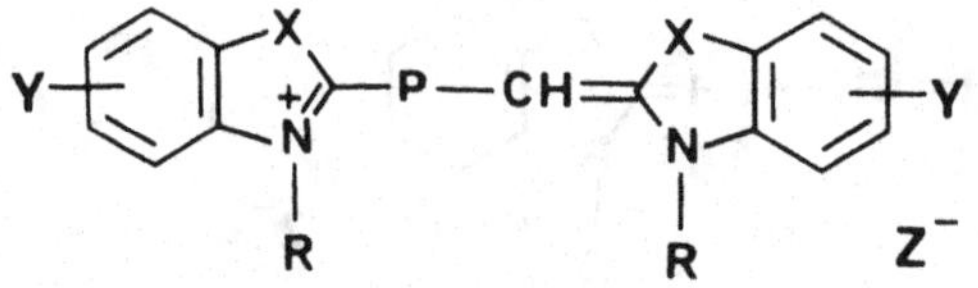

a. X = S, R = Et

Cyanine dyes	**(12)**	P = $-(-CH=CH-)_2$

Squarylium dyes   **(13)**   P =

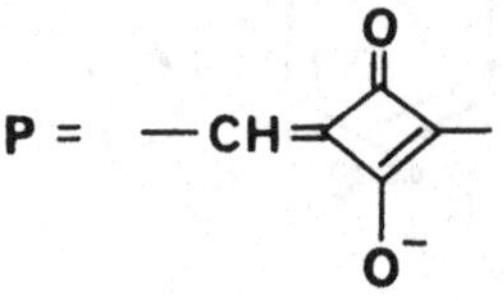

Croconium dyes   **(14)**   P =

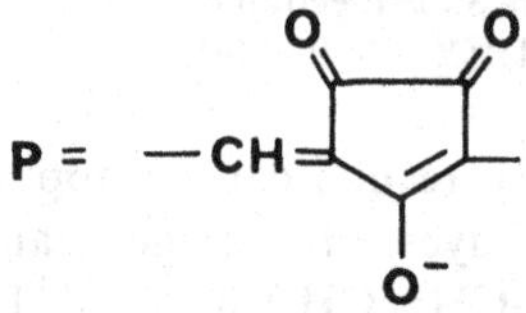

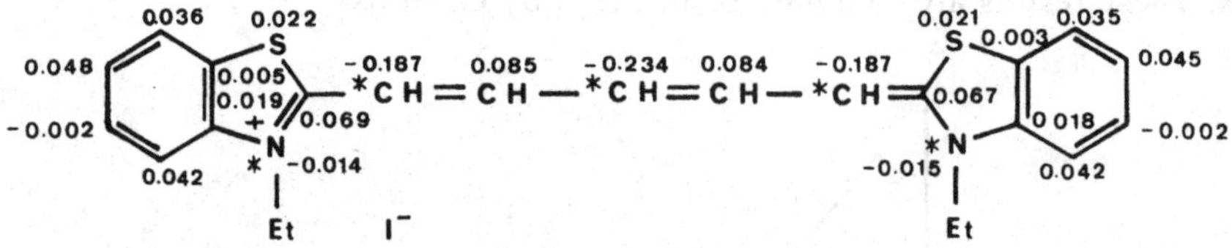

(12a)

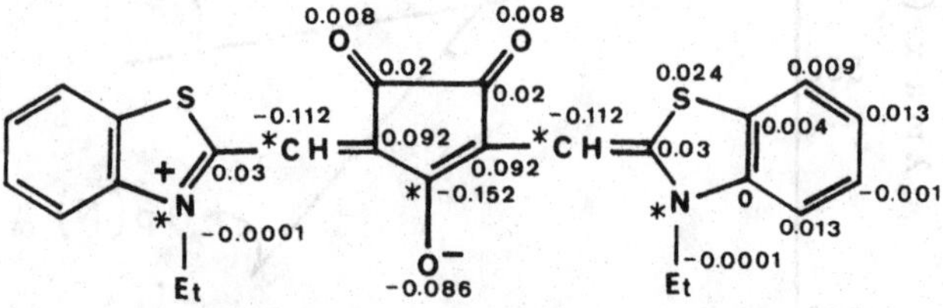

(14a)

Figure 5. π-Electron density changes accompanying the first excitation of dyes **12a** and **14a**.

The meso substituent effects of pentamethine cyanine dyes (**12**) on the λ_{max} of infrared absorbing cyanine dyes such as squarylium (**13**) and croconium dyes (**14**) have been studied recently by the PPP MO method.[17] The π-electron density changes accompanying the first excitation of dyes **12a** and **14a** are shown in Figure 5. The bathochromic shifts of **13a** (12 nm) and **14a** (120 nm) relative to **12a** are explained by the substituent effect at the meso positions; introduction of a carbonyl group (acceptor) at the unstarred position and that of an oxide group (donor) at the starred position of **12a** to give **13a** causes the bathochromic shift (Dewar's rule), and the introduction of a second carbonyl group at the unstarred position of **13a** gives **14a**, which absorbs in the infrared region. The absorption spectra of infrared absorbing cyanine dyes can be shifted quantitatively in this manner by applying the PPP MO method in the synthetic design of the dyes.

REFERENCES

1. C. R. Bury, *J. Am. Chem. Soc. 57*, 2116 (1935).
2. R. Pariser and R. G. Parr, *J. Chem. Phys. 21*, 466, 767 (1953); J. A. Pople, *Trans. Faraday Soc. 49*, 1375 (1953).
3. C. Lubai, C. Xing, H. Yufen, and J. Griffiths, *Dyes Pigments 10*, 123 (1989).
4. S. Tokita, M. Matsuoka, Y. Kogo, and H. Kihara, *Design of Functional Dyes by the PPP MO Method*, Maruzen, Tokyo (1989) (in Japanese).
5. J. Griffiths, *Colour and Constitution of Organic Molecules*, Academic Press, London (1976).
6. J. Fabian and H. Hartmann, *Light Absorption of Organic Colorants*, Springer-Verlag, Berlin (1980).
7. K. Y. Chu and J. Griffiths, *J. Chem. Soc. Perkin Trans. 1*, 696 (1979).
8. M. Matsuoka, K. Takagi, H. Obayashi, K. Wakasugi, and T. Kitao, *J. Soc. Dyers Colourists 99*, 257 (1983).
9. Y. Kogo, H. Kikuchi, M. Matsuoka, and T. Kitao, *J. Soc. Dyers Colourists 96*, 475 (1980).
10. S. H. Kim, M. Matsuoka, T. Yodoshi, K. Suga, and T. Kitao, *J. Soc. Dyers Colourists 105*, 212 (1989).
11. Y. Kubo, F. Mori, and K. Yoshida, *Chem. Lett.* 1761 (1987).
12. M. Matsuoka, T. Yodoshi, L. Han, and T. Kitao, *Dyes Pigments 9*, 343 (1988).
13. M. J. S. Dewar, *J. Chem. Soc.*, 2329 (1950); 3532, 3544 (1952).
14. E. B. Knott, *J. Chem. Soc.*, 1024 (1951).
15. M. Klessinger, *Theor. Chim. Acta (Berlin)*, *5*, 251 (1966).
16. S. Yasui, M. Matsuoka, and T. Kitao, *Shikizai Kyokaishi*, *61*, 375 (1988).
17. S. Yasui, M. Matsuoka, and T. Kitao, *Dyes Pigments 10*, 13 (1988).

3

Cyanine Dyes

MASARU MATSUOKA

I. INTRODUCTION

Polymethine cyanine dyes have the general structure 1 and absorb in a wide range of wavelengths, from 340 to 1400 nm. In structure 1, R denotes the heteroaromatic residue, and λ_{max} is affected strongly by the electronic characteristics of R. The length of an ethylene unit in the conjugating bridge also strongly affects λ_{max}, and near-infrared absorption generally can be attained for $n > 3$.

$$R \underset{\underset{R}{\overset{|}{N}}}{\overset{*}{C}} = CH - (CH = CH)_n - \overset{*}{C} \underset{\underset{R}{\overset{|}{N^+}}}{R}$$

1

The color-structure relationships of cyanines and related dyes have been quantitatively evaluated by the PPP MO method by Fabian and Hartmann.[1] The structural changes affecting the color of cyanine dyes have also been summarized by Griffiths.[2]

Over the past century, cyanine dyes have been used mainly as photosensitizers for silver halide photography, and several tens of thousands of

MASARU MATSUOKA • Department of Applied Chemistry, College of Engineering, University of Osaka Prefecture, Sakai, Osaka 591, Japan.

cyanine dyes have been developed, in keeping with the advance of technology in the photographic industry. The relationship between their photosensitizing ability and chemical structures has been established.[3]

On the other hand, there is much demand for the use of cyanine dyes as functional dyes for new technology. These new applications are listed in Table I.

In general, cyanine dyes have poor resistance to light compared with other dye chromophores such as azo and quinone dyes, but many structural modifications improving lightfastness and resistance can be made. Cyanine dyes have a wide variety of possible structural alternations and can be used as functional materials for high technology in numerous fields of application.

Syntheses and some new applications of cyanine dyes are described in this chapter. Applications of infrared absorbing cyanine dyes as photosensitizers for photography are described in Chapter 14 and are thus not included here. The basic chromophores of cyanine dyes have been examined thoroughly by Griffiths[2] and Fabian and Hartmann,[1] and a few are discussed in this chapter.

II. GENERAL SYNTHETIC METHOD

The syntheses of traditional and basic cyanine dyes have been summarized by Venkataraman.[4] Some examples of these are shown in Scheme 1.

Monomethine cyanine dye **2**, as an example, can be synthesized by the condensation reaction of the alkylammonium salts of 2-methylbenzoxazole and 2-methylthiobenzoxazole (Eq. 1). The trimethine cyanine dyes ($n = 1$) are also obtained as shown in Eq. (2).

The pentamethine cyanine dyes ($n = 2$) can be synthesized by a stepwise procedure as shown in Eq. (3). The asymmetrical cyanines are also obtainable by derivatizing the heteroaromatic residue as an active methylene

Table I. New Applications of Cyanine Dyes[a]

Photosensitizing dyes for xerography
Laser dyes
Functional dyes for LB films
Potential-sensitive dyes
Chromic dyes (photo-, electro-, thermo-, and piezochromic)
Pleochroic dyes for LC display
Infrared absorbing dyes for optical storage
Cosmetic ingredients and quasi-drugs

[a] Ref. 6.

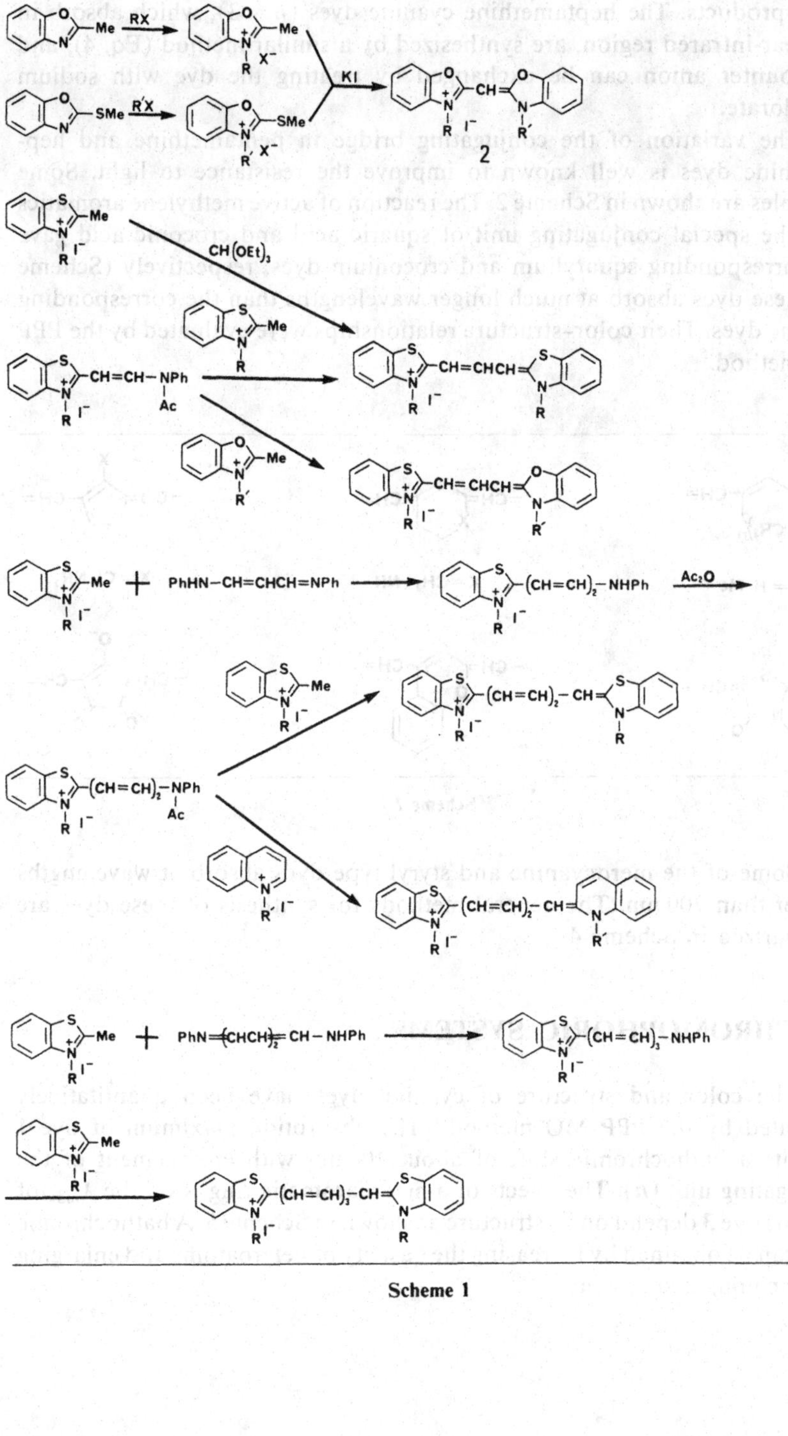

(1)

(2)

(3)

(4)

Scheme 1

compound. In these cases, some of the symmetrical cyanines are obtained as by-products. The heptamethine cyanine dyes ($n = 3$), which absorb in the near-infrared region, are synthesized by a similar method (Eq. 4), and the counter anion can be exchanged by treating the dye with sodium perchlorate.

The variation of the conjugating bridge in pentamethine and heptamethine dyes is well known to improve the resistance to light. Some examples are shown in Scheme 2. The reaction of active methylene aromatics with the special conjugating unit of squaric acid and croconic acid gave the corresponding squarylium and croconium dyes, respectively (Scheme 3). These dyes absorb at much longer wavelengths than the corresponding cyanine dyes. Their color–structure relationships were evaluated by the PPP MO method.[5]

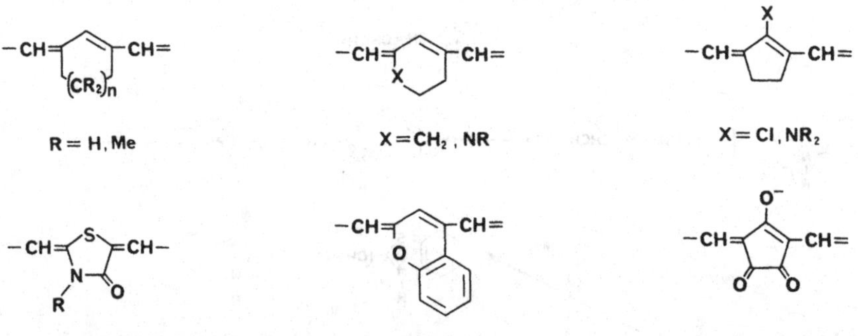

Scheme 2

Some of the merocyanine and styryl type dyes absorb at wavelengths greater than 700 nm. The general methods for synthesis of these dyes are summarized in Scheme 4.

III. CHROMOPHORIC SYSTEMS

The color and structure of cyanine dyes have been quantitatively evaluated by the PPP MO method.[1] The absorption maximum of dye **1** exhibits a bathochromic shift of about 100 nm with enlargement of the conjugating unit (n). The effects of a heteroaromatic ring R on the λ_{max} of cyanine dye **3** depend on its structure as shown in Scheme 5. A bathochromic shift can be obtained by increasing the basicity of heteroatoms and enlarging the π-conjugated systems.

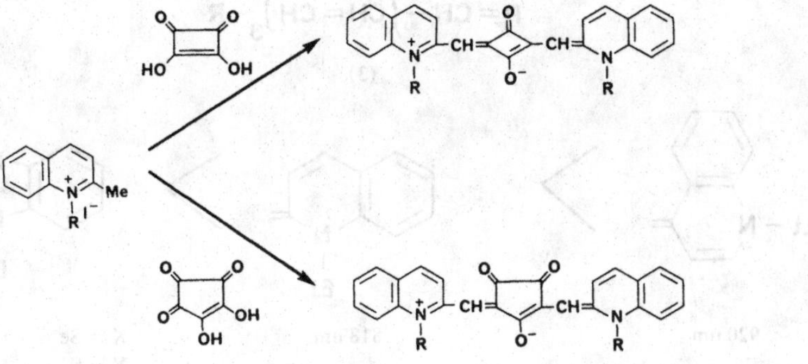

Scheme 3

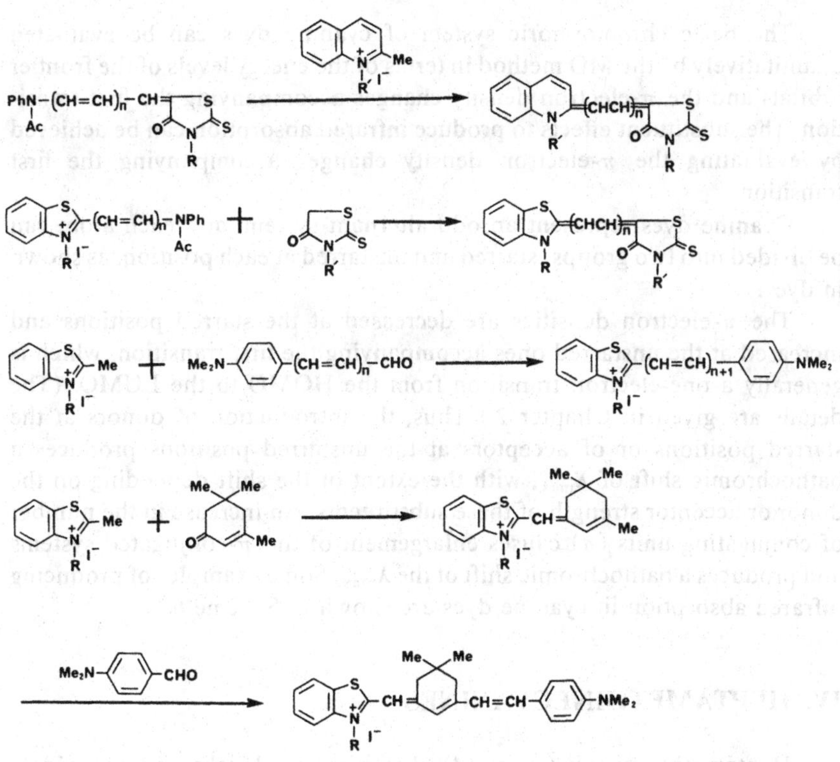

Scheme 4

$$\overset{+}{R} = CH - \left(CH = CH\right)_{\!3}^{} - R$$

(3)

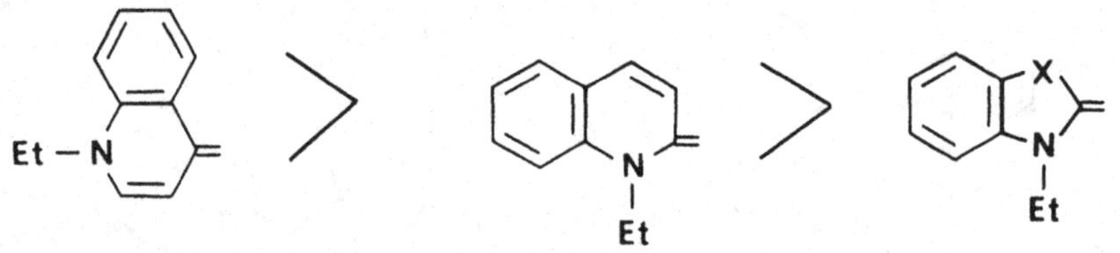

Et—N ...	> ... N	> ... N
		Et
920 nm	818 nm	

X = Se	790 nm
X = S	763 nm
X = CMe$_2$	741 nm
X = O	695 nm

Scheme 5

The basic chromophoric system of cyanine dyes can be evaluated quantitatively by the MO method in terms of the energy levels of the frontier orbitals and the π-electron density changes accompanying the first transition. The substituent effects to produce infrared absorption can be achieved by evaluating the π-electron density changes accompanying the first transition.

Cyanine dyes represent an odd alternant system, and each atom can be divided into two groups: starred and unstarred at each position, as shown in dye 1.

The π-electron densities are decreased at the starred positions and increased at the unstarred ones accompanying the first transition, which is generally a one-electron transition from the HOMO to the LUMO. (The details are given in Chapter 2.) Thus, the introduction of donors at the starred positions or of acceptors at the unstarred positions produces a bathochromic shift of λ_{max}, with the extent of the shift depending on the donor or acceptor strength of these substituents. An increase in the number of conjugating units (n) causes enlargement of the π-conjugated systems and produces a bathochromic shift of the λ_{max}. Some examples of producing infrared absorption in cyanine dyes are shown in Scheme 6.

IV. HEPTAMETHINECYANINES

Heptamethinecyanine dyes (4) absorb infrared light and are widely employed as light absorbing media in optical recording systems. As men-

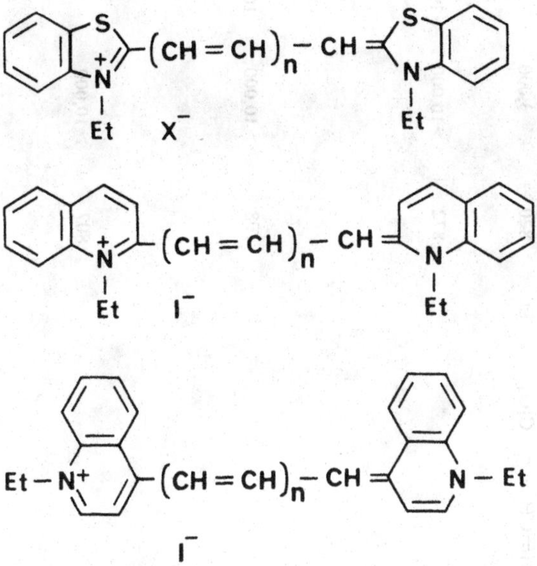

	λ_{max}
$n = 1$	557 nm
$n = 2$	653 nm
$n = 3$	757 nm
$n = 2$	708 nm
$n = 3$	818 nm
$n = 4$	943 nm
$n = 1$	710 nm
$n = 2$	810 nm
$n = 3$	920 nm

Scheme 6

tioned in Chapter 10, dye media must have good solubility in nonpolar organic solvents for the solvent-coating method. Cyanine dye has an N-alkyl substituent in the parent nucleus, and its solubility can be controlled by variation of the N-alkyl substituents. Some properties of cyanine dyes employed in optical memory disks are summarized in Table II. They absorb at 730–820 nm in solution, and their solubility in organic solvents is strongly affected by the nature of the N-alkyl substituent, the heteroaromatic ring, and the counter anion.

It is generally said that perchlorates dissolve much more in nonpolar solvents such as methylene chloride than in methanol. Indolenine derivatives have better solubility than benzothiazole or quinoline derivatives. The indolenine infrared absorbing dyes 4c and 4d have good properties such as solubility, heat resistance, and lightfastness and can be used as recording media.[6]

The N-alkyl substituent affects not only the solubility but also the aggregation of dye molecules, which lowers the performance of the recording layer. Longer alkyl chains in cyanine dyes prevent the crystallization caused by aggregation.

The lightfastness of cyanine dyes is generally poor, and structural modifications are required to improve it. This is the purpose of the replace-

Table II. Heptamethinecyanine (**4**) and related Dyes for Optical Recording Media[a]

	Y	X	R	λ_{max}[b] (nm)	Solubility[c]	
					CH$_3$OH	CH$_2$Cl$_2$
4a	S	I	Et	757	800	900
4b	S	ClO$_4$	Et	758	3,000	300
4c	C(Me)$_2$	I	Me	738	20	10
4d	C(Me)$_2$	ClO$_4$	Me	740	200	<10
4e	CH=CH	I	Et	817	1,000	150
4f	CH=CH	Br	Et	818	200	300
4g	CH=CH	ClO$_4$	Et	820	3,000	500
5				872	>10,000	>10,000
6				708	>10,000	10,000
7				807	>10,000	400

100

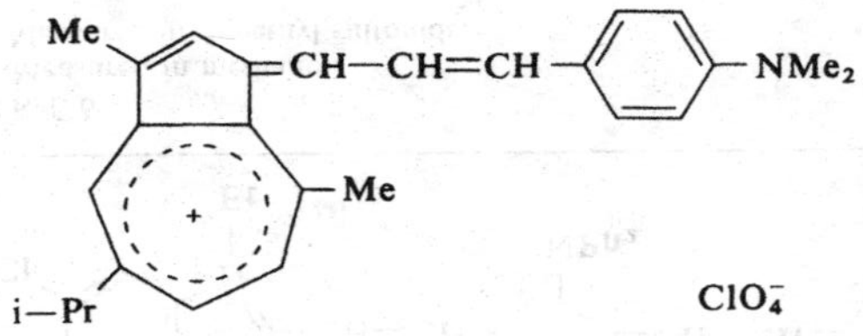

1,500

728[d]

8

ClO_4^-

NMe$_2$

CH—CH=CH

Me

Me

Me

i—Pr

[a] Ref. 6.
[b] Measured in methanol.
[c] Amount of solvent (ml) required to dissolve 1 g of dye.
[d] Measured in methylene chloride.

Table III. Heptamethinecyanine Infrared Absorbing Dyes for Dye Laser Materials[a]

Structure	X	Y	R	λ_{max} (nm)	L_{max} (nm)
	O	I	Me	681[b]	720–864
	C(Me)$_2$	I	Me	741[b]	775–940
	S	Br	Et	759[b]	793–900
				817[b]	865–969
				928[b]	970–1145
				825[c]	858–1030

[a] Ref. 6.
[b] Measured in methanol.
[c] Measured in dimethyl sulfoxide.

ment of an ethylene unit by a cyclic conjugated bridge in dye **5** and/or the introduction of a carbocyclic ring in the squarylium dye **6** and the croconium dye **7**. Styryl type azulenium dye **8** absorbs in the infrared region.

Quenchers such as nickel complexes are generally used to improve the lightfastness of infrared absorbing cyanine dyes. Cyanine dyes with a nickel complex as a counter anion are also patented; these have superior characteristics as practical recording media (cf. Chapter 6).

Polymethine infrared absorbing dyes are used as dye laser materials. The dye laser was developed by Sorokin in 1966, and the use of over 600 kinds of dyes in dye lasers has been reported. Dye lasers can emit a wide range of wavelengths from the ultraviolet to the infrared regions and can be used for spectrophotometric and medical applications (see Chapter 15).

Cyanine dyes are very useful as laser materials because of their high molar absorptivity, high fluorescence quantum yield, and large Stoke's shift, but they have the disadvantage of chemical instability and/or poor resistance to light.

Some of the heptamethinecyanine infrared laser dyes are listed in Table III[6] together with their spectral properties.

V. SQUARYLIUM AND CROCONIUM DYES

Synthesis of squarylium[7] and croconium dyes[8] has been reported. The squarylium dyes have an almost planar structure through all π-conjugated systems, from observations by X-ray analysis.[9,10] A chromophoric system for correlating their absorption spectra with their structures by means of the PPP MO method has recently been reported.[5] Effects of substituents on the λ_{max} of dyes **9-11** are summarized in Table IV. The introduction of a croconic moiety into the conjugated methine chains of cyanine dyes produces a 120–126 nm bathochromic shift; introduction of a squaric moiety produces only a **12–18** nm bathochromic shift in dyes **10a** and **10b**, while it produces a 9 nm hypsochromic shift in dye **10c**.

Thus, the croconium dyes **9** absorb at much longer wavelengths than the corresponding squarylium dyes **10** and cyanine dyes **11**. These results were confirmed by the PPP MO method. The halfwidth value at λ_{max} of these dyes increases in the order croconium < cyanine < squarylium. A narrow absorption band is important for use as dye laser material and recording media for laser optical storage.

The croconium and squarylium dyes can be evaluated as having the same chromophoric system as the corresponding cyanine dyes **11**. Dyes **9** can be considered to have two carbonyl groups as the acceptors at the unstarred positions and the oxide group as the donor at the starred position.

Table IV. Effect of Methine Substituents on the λ_{max} of Dyes 9, 10, and 11[a]

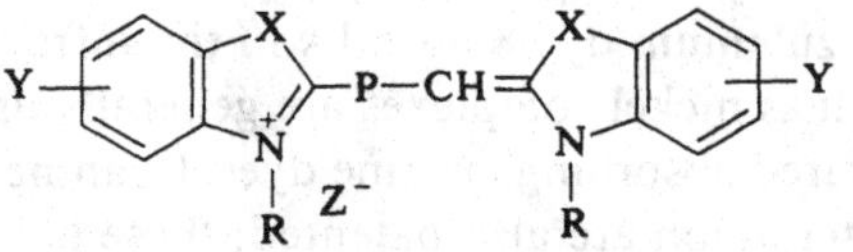

Croconium dyes (9)

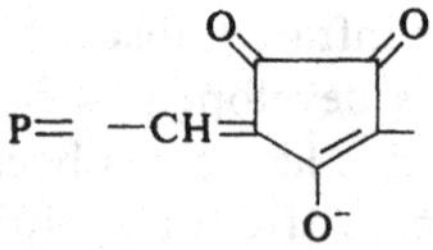

Squarylium dyes (10)

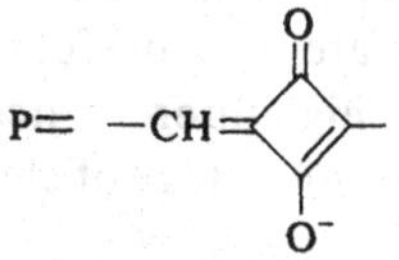

Cyanine dyes (11)   P= $-(CH=CH)_{\overline{x}}$

Dye	X	Y	R	Calculated λ_{max} (nm)	Observed[b] λ_{max} (nm)	$\Delta\lambda$ obs. − calc.	Observed $\Delta\lambda$ 9-10	9-11	10-11
9a	CH=CH	H	Et	731	832	101			
10a	CH=CH	H	Et	680	724	44	108		
11a	CH=CH	H	Et	671	706	35		126	18
9b	S	H	Et	709	771	62			
10b	S	H	Et	650	663	13	108		
11b	S	H	Et	617	651	34		120	12
9c	C(Me)$_2$	H	Me		764				
10c	C(Me)$_2$	H	Me		629		135		
11c	C(Me)$_2$	H	Me		638			126	−9

[a] Ref. 5.
[b] Measured in acetonitrile.

These substituents are expected to produce a large bathochromic shift of λ_{max} from its position in cyanine dyes 11. Squarylium dyes 10 have one carbonyl group at the unstarred position and an oxide group at the starred position. Thus, the croconium dyes 9 can be expected to absorb at much longer wavelengths than the squarylium dyes 10, which are expected to absorb at longer wavelengths than the cyanine dyes 11 (see Chapter 2, Figure 5). Some of these dyes are useful as dye media for optical recording systems and photosensitizers for electrophotography. Dye **9b** was reported to have good photosensitivity from the visible to the infrared region.[11] Dyes **9a–9c** and **10a** absorb above 700 nm and can be employed as optical recording media.

VI. PYRYLIUM AND RELATED DYES

Some of the pyrylium dyes absorb in the infrared region. Chalcogenopyrylomethine dyes, which have the general form **12**, absorb at much longer wavelengths than the corresponding cyanine dyes. The effects of heteroatom X on the λ_{max} of dye **12** follow the order Te (1010 nm) > Se (910) > O (798) > NR (748).

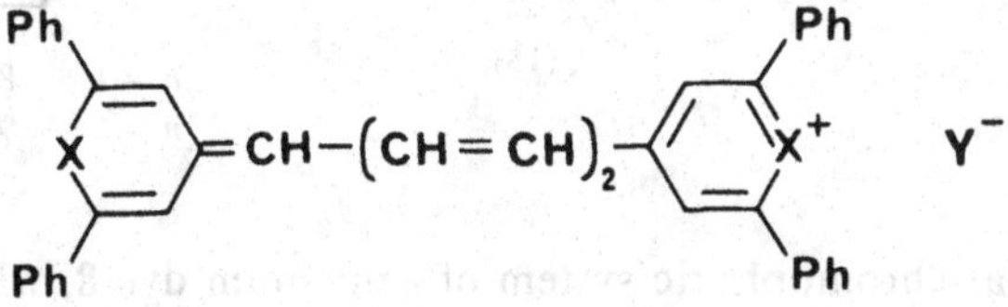

(12)

The monochalcogenopyrylomethine dyes **13** absorb at much shorter wavelengths than the corresponding chalcogenopyrylomethine dyes, but some absorb in the infrared region.

$n = 1$	790 nm, log ε 4.97
$n = 2$	870 nm, log ε 5.19

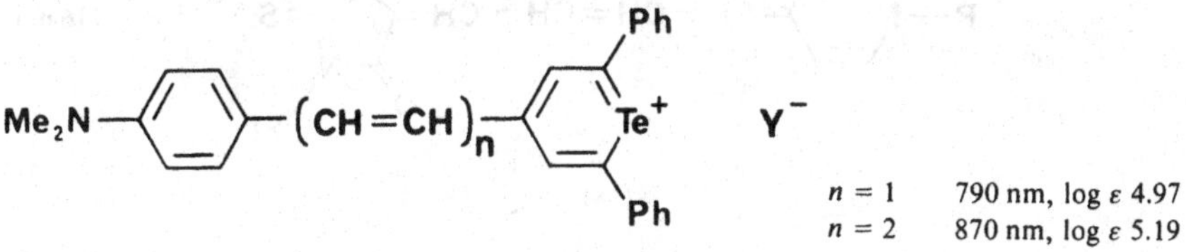

(13)

The similar pyrylium dyes of triphenylmethane ethynologues **14**[12] and bis-substituted analogues **15**[13] were recently reported.

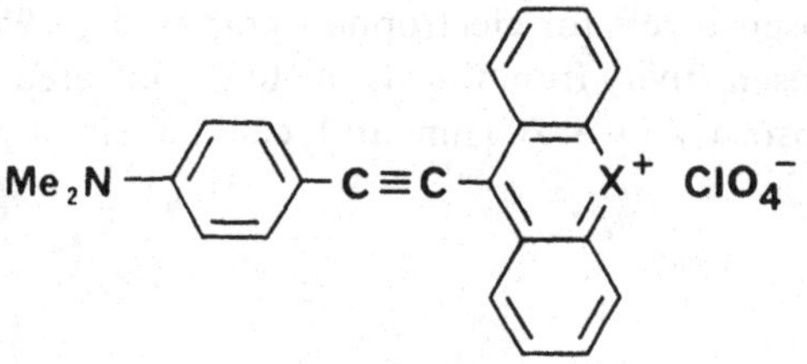

$$\lambda_{max} \text{ 730–800 nm}$$
$$X = O, S, Se$$

(14)

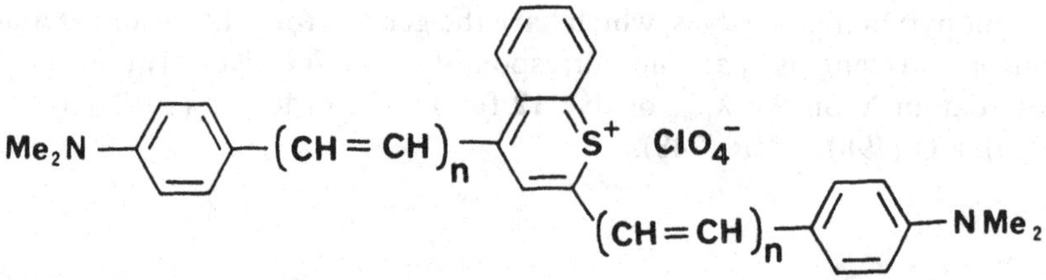

(15)

$n = 1$	836 nm (10, 300)
$n = 2$	952 nm (52,200)

The similar chromophoric system of azulenium dye **8**, which absorbs at 728 nm, has been reported as the charge-generating material (CGM) in organic photoconductors for laser printers and/or optical recording media.

Some merocyanine dyes of general formula **16** absorb in the infrared region. The first transition consists of a strong intramolecular charge transfer.

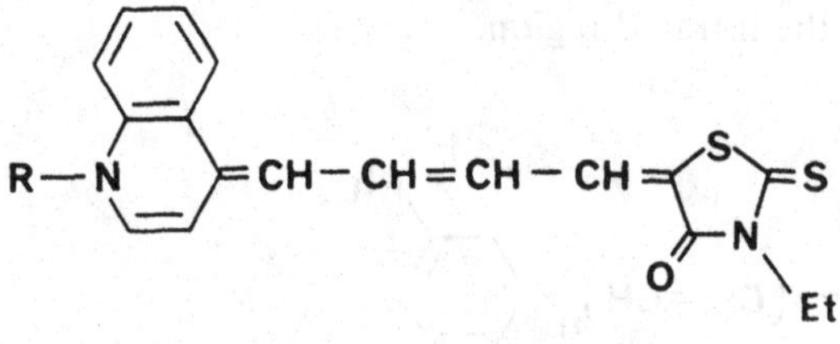

716 nm

(16)

The indoline type cyanine dye **17** was reported as an electrochromic dye.[14] The color–colorless system of dye **17** is the same as that of spiropyran (see Chapter 7). Some cyanine and merocyanine infrared absorbing dyes are used for medical applications and as potential-sensitive dyes for cell membranes.[6]

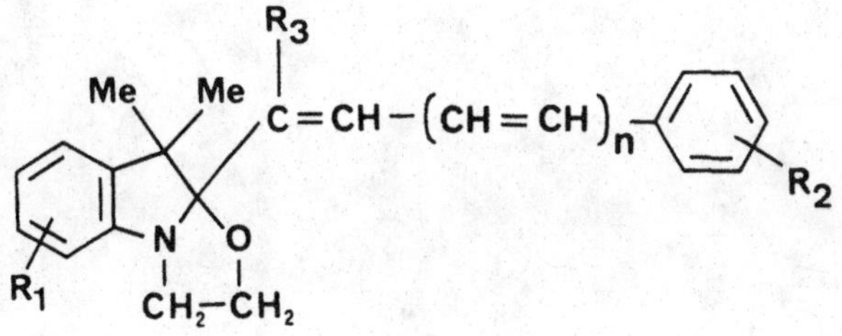

(17)

Colorless

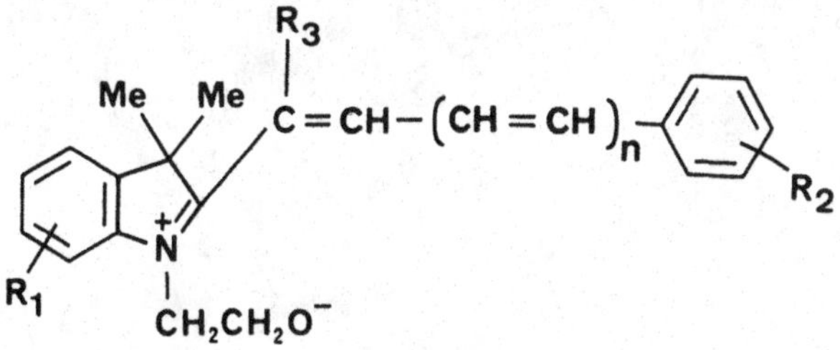

‖

Colored

REFERENCES

1. J. Fabian and H. Hartmann, *Light Absorption of Organic Colorants*, Springer-Verlag, Berlin (1980).
2. J. Griffiths, *Color and Constitution of Organic Molecules*, Academic Press, London (1976).
3. T. H. James, *The Theory of the Photographic Process*, 4th ed., Macmillan, New York (1977).
4. K. Venkataraman, in: *The Chemistry of Synthetic Dyes* (K. Venkataraman, ed.), Vol. II, p. 1143, Vol. IV, p. 212, Academic Press, London (1952).
5. S. Yasui, M. Matsuoka, and T. Kitao, *Dyes Pigments 10*, 13 (1988).
6. S. Yasui, *Shikizai Kyokaishi 60*, 212 (1987).
7. V. H. E. Sprenger and W. Ziegenbein, *Angew. Chem. 78*, 581 (1967).
8. G. Alfred, Ger. Offen. 1930224 (1970).
9. Y. Kobayashi, M. Koto, and M. Kurahashi, *Bull. Chem. Soc. Jpn. 59*, 311 (1986).
10. J. Bernstein, M. T. Kendra, and C. J. Kekhardt, *J. Phys. Chem. 90*, 1069 (1986).
11. G. A. Rillaers and H. M. Depoorter, Jpn. Patent 51-41061.
12. S. Akiyama, S. Nakatsuji, K. Nakashima, M. Watanabe, *J. Chem. Soc. Chem. Commun.*, 710 (1987).
13. H. Nakazumi, S. Watanabe, S. Kado, and T. Kitao, *Chem. Lett.*, 1039 (1989).
14. M. Hayami and S. Torigoshi, Jpn. Patent 1,158,538, 61-25036.

4
Quinone Dyes

MASARU MATSUOKA

I. INTRODUCTION

Dyes based on the quinone chromophore are widely used in the modern dyestuff industry, and a wide range of colorants have been commercialized. The chemistry of quinone dyes was reviewed thoroughly by Venkataraman[1] in 1971. Gordon and Gregory[2] also reviewed advances in anthraquinone dye chemistry in 1983, but they did not describe the infrared absorbing quinone dyes.

Quinone molecules are well known as electron acceptors and emit virtually no light in the visible region. The introduction of donors and/or acceptors into the quinone nucleus produces visible absorption with intramolecular charge-transfer character. 1,4-Naphthoquinone and 9,10-anthraquinone are the main chromophores for the colorants. The 1,4-naphthoquinones, which have not been used as commercial dyes because of their instability and difficult synthesis compared with 9,10-anthraquinones, are particularly interesting as deep-colored quinone dyes. It has been found that donor-substituted naphthoquinones absorb at much longer wavelengths than the corresponding anthraquinones. Development of dyes which are small in molecular size but produce deep color is of interest because of their potential use in guest–host liquid-crystal displays, dye diffusion thermal transfer systems, optical recording systems, and so on. The synthesis and characterization of infrared absorbing quinone dyes

MASARU MATSUOKA • Department of Applied Chemistry, College of Engineering, University of Osaka Prefecture, Sakai, Osaka 591, Japan.

are currently the most interesting and important subjects in terms of optical recording media. The development of organic photoconductors which absorb the near-infrared light of produced by semiconductor lasers is also very important for high-speed laser printer systems. In this section, the new chemistry of quinone infrared absorbing dyes is described. This includes the synthesis and characterization of 1,4-naphthoquinones, phenothiazinequinones, indonaphthols, anthraquinones, and other quinone type dyes as well some applications of these dyes in high technology.

II. SYNTHESIS AND CHARACTERISTICS OF INFRARED ABSORBING QUINONE DYES

The first synthesis of near-infrared absorbing 1,4-naphthoquinone dyes was reported by Chu and Griffiths[3] in 1978. They obtained 5-amino-8-anilino-2,3-dicyano-1,4-naphthoquinone, which has a λ_{max} at 759 nm in acetone, by the reaction of 5-amino-2,3-dicyano-1,4-naphthoquinone (1) with aniline in ethanol. Little attention was drawn to its qualities as an infrared absorbing dye at that time, but the novel direct 8-arylamination of 1 was reexamined by Kasai and co-workers[4] in 1981. They proposed that 8-arylamination was initiated by the formation of a π complex between 1 and the arylamine and that this π complex was transformed to the σ complex, which was then oxidized by 1 to give 2. However, the mechanism of the direct 8-arylamination of 1 was not known. The details of this reaction were reported by Matsuoka and co-workers[5] in 1985. The initial quinone-quinoneimine tautomerism of 1 to 4-hydroxy-2,3-dicyano-5-imino-1,5-naphthoquinone facilitates the 8-arylamination of 1. Matsuoka and co-workers isolated all the reaction products and explained the reactivity by the results of PPP MO calculations. 5,8-Diarylamino- (3) and 5-amino-8-hydroxy-2,3-dicyano-1,4-naphthoquinones (4), which also absorbed in the near-infrared region, were isolated as the by-products (Scheme 1).

The quinone–quinoneimine tautomerism of dye 4 in solution was observed; 4 absorbed at 583 nm in quinoneimine form and at 754 nm in quinone form.[6] The isosbestic point is observed at 660 nm in a mixture of chloroform and dimethyl sulfoxide. These tautomerisms of quinone dyes are very interesting in the molecular design of infrared absorbing quinone type dyes.

Matsuoka and co-workers[7] reported the reaction of 2,3-dichloro-5,8-dihydroxy-1,4-naphthoquinone (2,3-dichloronaphthazarin) (5) with potassium 2-aminobenzenethiolate (6c) to give a new type of deep-colored quinone dye. The reactions are described in Scheme 2. The reaction of 5 with 2-aminophenol (6a) or 1,2-diaminobenzene (6b) gave only the mono-

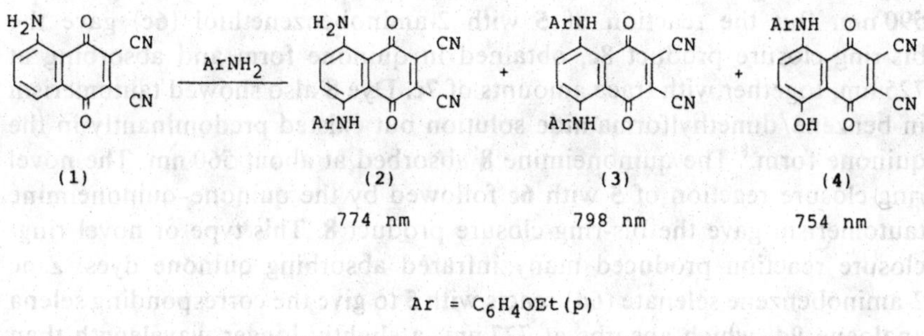

Ar = C$_6$H$_4$OEt(p)

Scheme 1

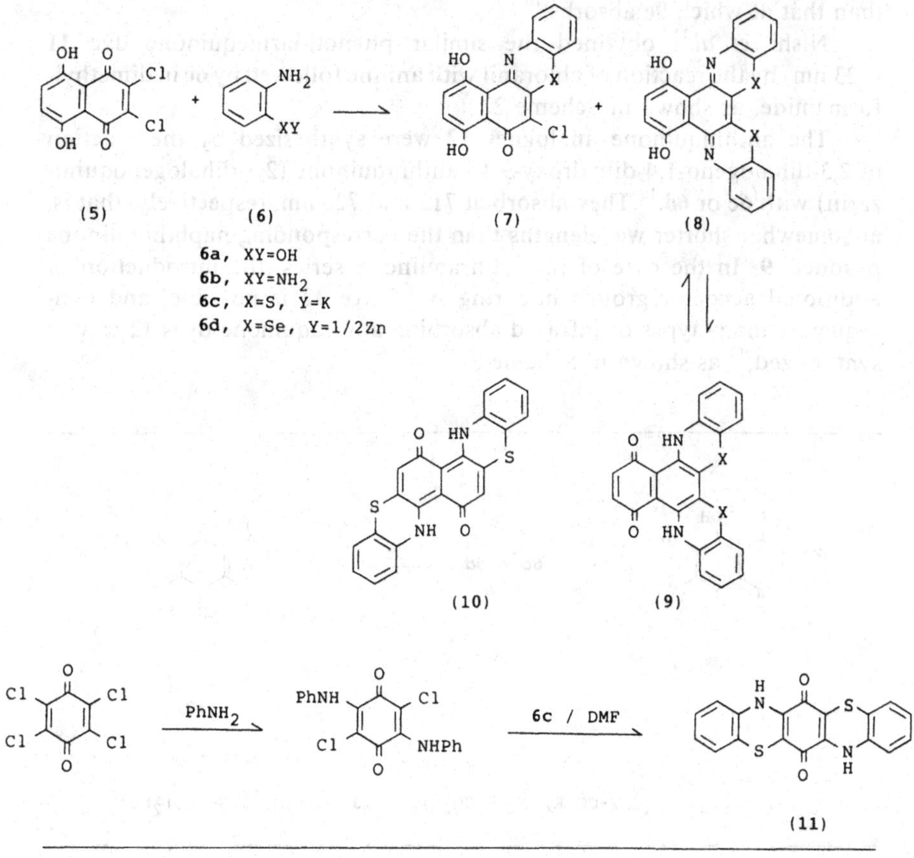

6a, XY=OH
6b, XY=NH$_2$
6c, X=S, Y=K
6d, X=Se, Y=1/2Zn

Scheme 2

ring-closure product 7 in quinoneimine form, which absorbed at 580–590 nm. But the reaction of 5 with 2-aminobenzenethiol (6c) gave the bis-ring-closure product 8c, obtained in quinone form and absorbing at 725 nm, together with trace amounts of 7c. Dye 8 also showed tautomerism in benzene/dimethylformamide solution but existed predominantly in the quinone form.[8] The quinoneimine 8 absorbed at about 560 nm. The novel ring-closure reaction of 5 with 6c followed by the quinone–quinoneimine tautomerism gave the bis-ring-closure product 8. This type of novel ring-closure reaction produced many infrared absorbing quinone dyes. Zinc 2-aminobenzene selenate (6d) reacts with 5 to give the corresponding selena analogue 9d, which absorbs at 727 nm, a slightly longer wavelength than 725 nm at which 9c absorbs.[9] A similar reaction of tetrabromonaphthazarin with 6c gives the corresponding 1,5-bis-ring-closure product 10, which absorbs infrared light at 780 nm, that is, at a wavelength considerably longer than that at which 9c absorbs.[10]

Nishi et al.[11] obtained the similar phenothiazinequinone dye 11 (723 nm) by the reaction of chloranil with aniline followed by 6c in dimethylformamide, as shown in Scheme 2.

The anthraquinone analogues 12 were synthesized by the reaction of 2,3-dihalogeno-1,4-dihydroxy-9,10-anthraquinone (2,3-dihalogenoquinizarin) with 6c or 6d.[12] They absorb at 712 and 720 nm, respectively, that is, at somewhat shorter wavelengths than the corresponding naphthoquinone products 9. In the case of the anthraquinone series, the introduction of additional acceptor groups into ring A of dye 12 is possible, and consequently many types of infrared absorbing anthraquinone dyes 12 can be synthesized,[13] as shown in Scheme 3.

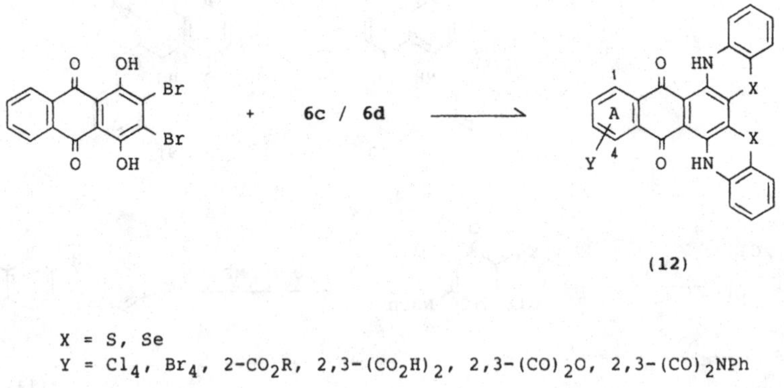

(12)

X = S, Se
Y = Cl_4, Br_4, 2-CO_2R, 2,3-$(CO_2H)_2$, 2,3-$(CO)_2O$, 2,3-$(CO)_2NPh$

Scheme 3

Indonaphthol dyes are well known for their cyan color in photography. They were studied widely by Weissberger and co-workers[14] in 1961. From the results of molecular design by the PPP MO method as described in Chapter 2, the introduction of acceptors into the quinone moiety and/or that of donors into the aniline moiety is expected to produce a bathochromic shift of λ_{max}.[15] Introduction of 2-carboxyamide produced a bathochromic shift of about 120–150 nm, and the resulting dyes absorb at 690–730 nm. Matsuoka and co-workers[16] reported that the dicyano derivative **13b**, obtained by cyanation of a monochloro derivative of **13a**, absorbs at 795 nm. A 212-nm bathochromic shift of λ_{max} is associated with the introduction of a dicyano group into the quinone moiety. The dye **13b** is very unstable and gradually reduces to the corresponding leuco compound, which no longer absorbs in the near-infrared region (Scheme 4).

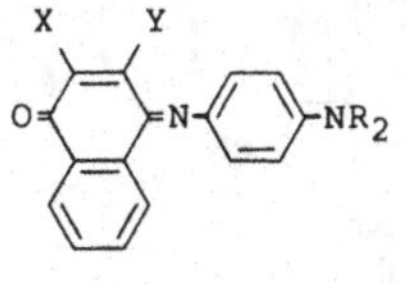

(13)

13a, X=H, Y=Cl or X=Cl, Y=H
13b, X=Y=CN

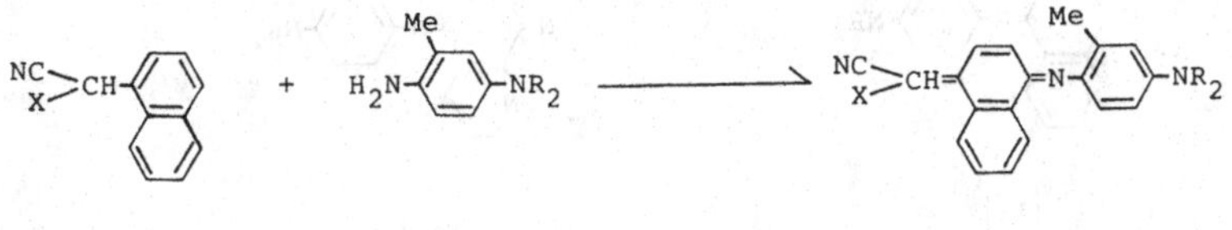

(14)

14a, X=CN
14b, X=CONH$_2$

Scheme 4

In 1987 Yoshida and co-workers[17-19] synthesized cyanomethide derivatives of indonaphthol (**14**) absorbing at 720–760 nm in chloroform by the reaction of 1-naphthylmalononitrile or 1-naphthylcyanoacetamide with *p*-*N,N*-dialkylaminoaniline in the presence of an oxidizing agent under alkali conditions (Scheme 4). The strong intramolecular charge-transfer character of the first excitation was confirmed by PPP MO calculation analysis of the chromophores. The cyanomethide group plays a major role as a strong acceptor in the indonaphthol chromophore. Yoshida and co-workers recently reported the synthesis of a series of quinoid ligands to give infrared absorbing metal complex dyes (Scheme 5). These are 5,8-quinolinediones (15),[20] 1,2-naphthoquinones (16),[21] 1-aza-9,10-anthraquinones (17),[22] azaindonaphthols (18),[23,24] and 3-phenyliminopyrido[2,3-a]phenothiazines (19).[25] All of these absorb in the visible region as free ligands but absorb in the infrared region after the formation of a 1:1 or 2:1 (metal:ligand) metal complex with Ni^{2+} or Cu^{2+} ions. Bathochromic shifts of 30–250 nm in λ_{max}, and the ε values attained upon metal complex formation are 1–10 times those of the free ligands. These are new types of infrared absorbing dyes and their characteristics are under investigation.

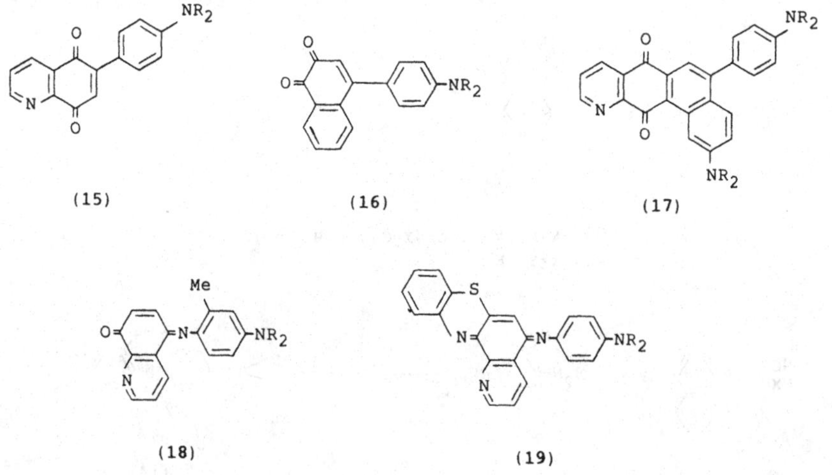

(15)　　　　　　　　(16)　　　　　　　　(17)

(18)　　　　　　　　(19)

Scheme 5

Some anthraquinone derivatives absorb above 700 nm. These include the 1,4-diaminoanthraquinone(N-alkyl)-3'-thioxo-2,3-dicarboximides (20), employed as deep-colored dichroic dyes for guest–host liquid-crystal displays,[26] the well-known indanthrene pigments (21),[27] employed as optical recording media, and 2-arylamino-3,4-phthaloylacridones (22)[28] for deep-colored disperse dyes or pigments. The novel quinone methide chromophore trisphenoquinone (23) also absorbs strongly at 769 nm (Scheme 6).

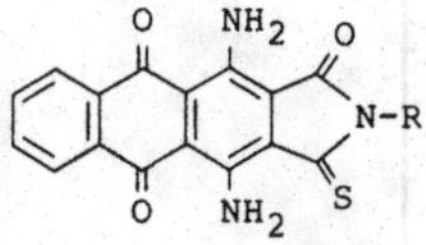

(20)

760 nm

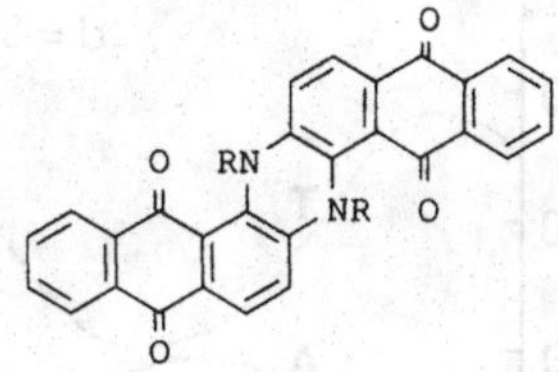

(21)

720 - 790 nm

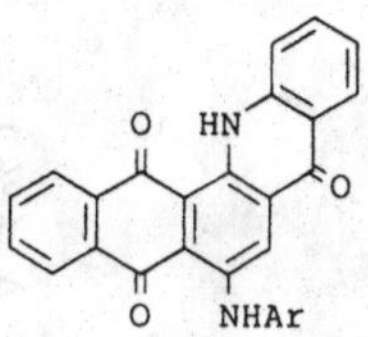

(22)

710 - 730 nm

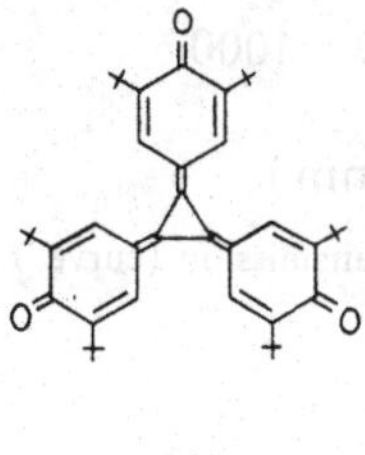

(23)

770 nm

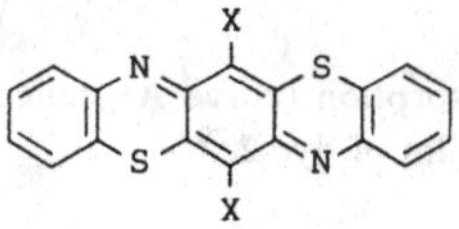

(24)

X=H, Cl

Scheme 6

III. APPLICATIONS OF INFRARED ABSORBING QUINONE DYES

Quinoid infrared absorbing dyes have been evaluated as dye media for optical recording systems. Some infrared absorbing naphthoquinone dyes **2** have good properties as recording media for optical recording systems. Many derivatives of **2** were synthesized and their characteristics as recording media for diode-laser high-density recording systems reported by Itoh *et al.*[29] in 1983.

It has been confirmed that the near-infrared absorbing dye medium composed of **2** has the following excellent advantages: high sensitivity for diode lasers due to high optical absorption and low melting/decomposition temperature characteristics, high signal-to-noise ratio due to clean shape of holes, long lifetime, and wide thickness latitude due to simple medium structure. The optical characteristics of dye film composed of **2** were also reported (Figure 1). The chemical modifications of **2** into the dimer form have been patented.

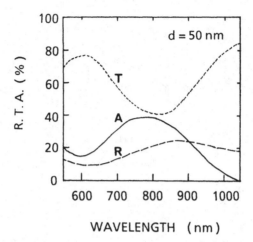

Figure 1. Absorption (curve A), reflection (curve R) and transmission (curve T) spectra of 50-nm-thick film of dye **2**.[29]

Naphthoquinone methide infrared absorbing dyes[17] are candidates for use as optical recording media. They have sufficient solubility in nonpolar organic solvents and can be applied by the solvent coating method onto polymer substrate to prepare dye media and thus have many advantages for practical use. A film of **14a** exhibited a broad absorption peak at a wavelength of 600–1000 nm and a λ_{max} value of 785 nm. The film reflected 23.3% of the incident light intensity at 830 nm and was found to be very suitable as a diode-laser optical storage medium.

Nickel-complex dyes **25** prepared from ligand **18**[23,24] have sufficient solubility in tetrachloroethane and can be applied by the solvent coating method to prepare dye media. The film absorbed strongly at 800 nm and reflected over 40% of the incident light intensity at 830 nm. These dyes have very large ε values, of over 100,000, like cyanine type dyes, and are interesting for their characteristics as optical recording media. This type of dye is described in Chapter 6.

$$\left[\text{Ni} \cdots \underset{N}{\overset{O}{\bigcirc}} = N - \bigcirc\!\!-\!\!\underset{Me}{} - NEt_2 \right]^{2+} ClO_4^{2-} \cdot 2$$

(25)

Nishi and co-workers[30] reported the preparation of a thin film of triphenodithiazine dye **24** (Scheme 6) which absorbed in the infrared region on exposure to gases such as nitrogen oxide. The spectral properties in the solid state and the conductivity of **24** were studied, focusing on the application of this dye in chemical sensors.

REFERENCES

1. K. Venkataraman, in: *The Chemistry of Synthetic Dyes* (K. Venkataraman, ed.), Vol. V, p. 132, Academic Press, London (1971).
2. P. F. Gordon and P. Gregory, *Organic Chemistry in Colour*, p. 163, Springer-Verlag, London (1983).
3. K. Y. Chu and J. Griffiths, *J. Chem. Res.*, (S), 180, (M), 2319 (1978).
4. T. Nakamori, T. Chiba, and T. Kasai, *Nippon Kagaku Kaishi*, 1916 (1981).
5. K. Takagi, M. Matsuoka, Y. Kubo, and T. Kitao, *Dyes Pigments 6*, 75 (1985).
6. K. Takagi, M. Matsuoka, Y. Kubo, and T. Kitao, *J. Soc. Dyers Colourists 101*, 140 (1985).
7. K. Takagi, M. Kawabe, M. Matsuoka, and T. Kitao, *Dyes Pigments 6*, 177 (1985).
8. M. Matsuoka, S. H. Kim, Y. Kubo, and T. Kitao, *J. Soc. Dyers Colourists 102*, 232 (1986).
9. S. H. Kim, M. Matsuoka, and T. Kitao, *Chem. Lett.*, 1351 (1985).
10. K. Takagi, S. Kanamoto, K. Itoh, M. Matsuoka, S. H. Kim, and T. Kitao, *Dyes Pigments 8*, 71 (1987).
11. H. Nishi, Y. Hatada, and K. Kitahara, *Bull. Chem. Soc. Jpn. 56*, 1482 (1983).
12. S. H. Kim, M. Matsuoka, Y. Kubo, T. Yodoshi, and T. Kitao, *Dyes Pigments 7*, 93 (1986).
13. S. H. Kim, M. Matsuoka, T. Yodoshi, and T. Kitao, *Chem. Express 1*, 129 (1986).
14. C. R. Barr, G. H. Brown, J. R. Thirtle, and A. Weissberger, *Photogr. Sci. Eng. 5*, 195 (1961).
15. M. Matsuoka, in: *Hikarikiroku Gijutsu to Zairyo*, p. 176, CMC Inc., Tokyo (1985) (in Japanese).
16. S. H. Kim, M. Matsuoka, T. Yodoshi, K. Suga, and T. Kitao, *J. Soc. Dyers Colourists 104*, 212 (1989).
17. Y. Kubo, F. Mori, and K. Yoshida, *Chem. Lett.*, 1761 (1987).
18. Y. Kubo, F. Mori, K. Komatsu, and K. Yoshida, *J. Chem. Soc. Perkin Trans. 1*, 2439 (1988).
19. Y. Kubo, M. Kuwana, K. Yoshida, Y. Tomotake, T. Matsuzaki, and S. Maeda, *J. Chem. Soc. Chem. Commun.*, 35, (1989).
20. K. Yoshida, M. Ishiguro, and Y. Kubo, *Chem. Lett.*, 2057 (1987).
21. K. Yoshida, T. Kougiri, N. Oga, M. Ishiguro, and Y. Kubo, *J. Chem. Soc. Chem. Commun.*, 708 (1989).
22. K. Yoshida, T. Kougiri, E. Sakamoto, and Y. Kubo, *Bull. Chem. Soc. Jpn.*, in press.
23. Y. Kubo, K. Sasaki, and K. Yoshida, *Chem. Lett.*, 1563 (1987).
24. Y. Kubo, K. Sasaki, H. Kataoka, and K. Yoshida, *J. Chem. Soc. Perkin Trans. 1*, 1469 (1989).
25. Y. Kubo, H. Kataoka, and K. Yoshida, *J. Chem. Soc. Chem. Commun.*, 1457 (1988).
26. Nippon Kayaku Co. Ltd., *Jpn. Patent* 58-219262.
27. Ricoh Co. Ltd., Jpn. Patent 58-169152, 58-22448.
28. I. L. Boguslavskaya and V. I. Gudzenko, *J. Org. Chem. USSR, 4*, 96 (1968).
29. M. Itoh, S. Esho, K. Nakagawa, and M. Matsuoka, *Proc. SPIE-Int. Soc. Opt. Eng. 420*, 332 (1983).
30. I. Shirotani, N. Sato, H. Nishi, K. Fukuhara, T. Kajiwara, and H. Inokuchi, *Nippon Kagaku Kaishi*, 485 (1986).

5

Phthalocyanine and Naphthalocyanine Dyes

MASARU MATSUOKA

I. INTRODUCTION

Phthalocyanine chromophores were synthesized by chance in 1928 during the preparation of phthalimide from phthalic anhydride and ammonia in a reaction vessel made of iron. The iron phthalocyanine obtained was isolated and identified by Linstead[1] in 1934. Its structure was confirmed by Robertson[2] using X-ray analysis in 1935. Since then, phthalocyanines have become important colorants as dyes and pigments. Their structural analogy to the natural pigments such as the porphyrins is of great interest in academic research and also in regard to their applications as colorants. The complexes formed between transition metals, especially copper, and phthalocyanine are chemically very stable to light and heat, and are widely applied as commercial dyes and pigments. The chemistry of phthalocyanine compounds was reviewed by Moser and Thomas[3] in 1963, and further advances in the color chemistry of phthalocyanines were reviewed by Booth[4] in 1971 and Gordon and Gregory[5] in 1983.

In the search for infrared absorbing dyes for optical recording media, phthalocyanines (1; Scheme 1) were the first candidates and have been evaluated widely. Metal-free phthalocyanine absorbs at 698 nm in 1-chloronaphthalene and 772 nm in the solid state. The metal complexes

---

MASARU MATSUOKA • Department of Applied Chemistry, College of Engineering, University of Osaka Prefecture, Sakai, Osaka 591, Japan.

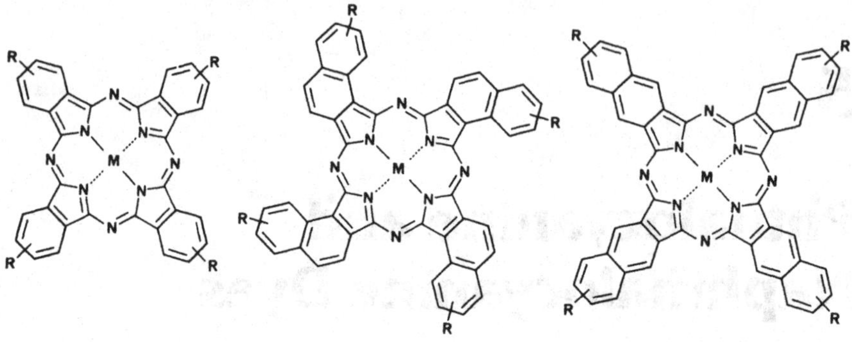

(1)   (2)   (3)

Scheme 1

generally absorb at much shorter wavelengths, but some, such as lead phthalocyanine, absorb at much longer wavelengths than metal-free phthalocyanine.[6]

On the other hand, the annelation of a phenyl ring of 1 to give 1,2-naphthalocyanines (2) and 2,3-naphthalocyanines (3) produces a large bathochromic shift of the λ_{max}. A series of 2,3-naphthalocyanine derivatives are potentially very important as organic materials for electro-optical applications such as optical recording media, organic photoconductors, color filter dyes, and photosensors. They can be modified to get improved solubility by the introduction of branched long-chain alkyl groups into the naphthalene rings and/or of trialkylsiloxysilane into the central core.

In this chapter, syntheses, color–structure relationships, and characteristics of phthalocyanines and naphthalocyanines are described.

II. SYNTHETIC METHOD

There are many practical synthetic methods for the preparation of phthalocyanines but, in principle, four main processes are generally employed (Scheme 2). In method 1, phthalic anhydride is heated with urea, a metal salt, and a catalytic amount of ammonium molybdate in a high-boiling solvent. Phthalic anhydride is converted by the ammonia liberated from the urea to phthalimide, monoiminophthalimide, and then diiminoisoindoline, which spontaneously tetramerizes and is oxidized to phthalocyanine. In method 2, phthalonitrile is heated in the presence of a base, a metal salt, and ammonium acetate with or without a solvent. Method

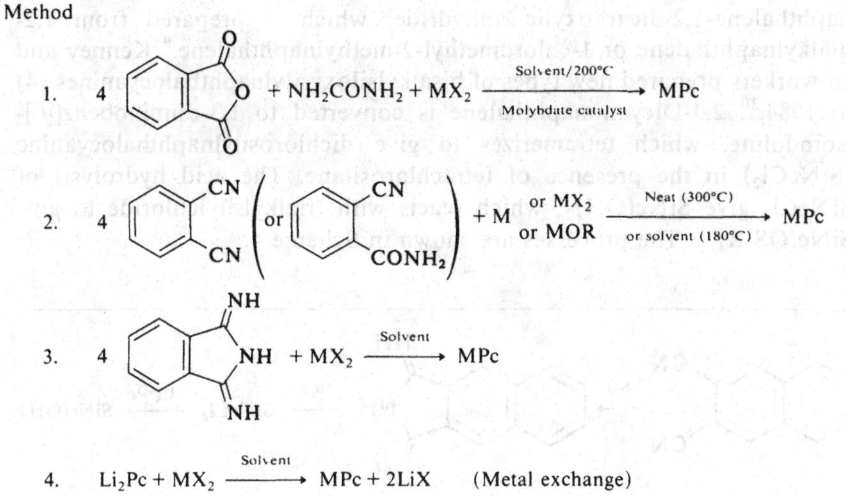

Method

1. 4 [structure: phthalic anhydride] $+ NH_2CONH_2 + MX_2$ $\xrightarrow[\text{Molybdate catalyst}]{\text{Solvent/200°C}}$ MPc

2. 4 $\left(\text{[structure: benzene with CN, CN]} \quad \text{or} \quad \text{[structure: benzene with CN, CONH}_2\text{]}\right)$ $+ M$ $\begin{array}{c}\text{or } MX_2 \\ \text{or } MOR\end{array}$ $\xrightarrow[\text{or solvent (180°C)}]{\text{Neat (300°C)}}$ MPc

3. 4 [structure: 1,3-diiminoisoindoline with NH, NH, NH] $+ MX_2$ $\xrightarrow{\text{Solvent}}$ MPc

4. $Li_2Pc + MX_2$ $\xrightarrow{\text{Solvent}}$ $MPc + 2LiX$ (Metal exchange)

Pc = phthalocyanine, M = metal

Scheme 2

3 is a variant of method 1 in which the isolated 1,3-diiminoisoindoline is used as the starting material. Method 4 is the metal exchange reaction of a labile metal phthalocyanine.

These four methods can be applied in the synthesis of naph-thalocyanines. 2,3-Naphthalocyanine was first synthesized by Luk'yanets and co-workers[7] in 1971 from 2,3-dicyanonaphthalene. The substituted naphthalocyanines are generally synthesized from substituted di-cyanonaphthalenes, which are synthesized as shown in Scheme 3. 2-3-Dicyanonaphthalene can be prepared by the direct ammoxidation of naphthalene-2,3-dicarboxylic anhydride, which is the by-product of the industrial process for the manufacture of anthraquinone from anthracene.[8]

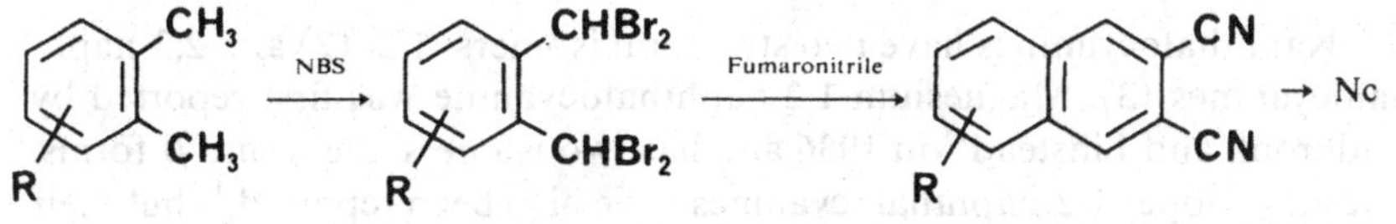

[structure: benzene with CH₃, CH₃, R] $\xrightarrow{\text{NBS}}$ [structure: benzene with CHBr₂, CHBr₂, R] $\xrightarrow{\text{Fumaronitrile}}$ [structure: naphthalene with CN, CN, R] $\rightarrow$ Nc

R = SiR₃, OR, SO₂NR₂, Ph, CO₂R, C$_n$H$_{2n+1}$, O(CH₂)$_n$OR, etc.

Scheme 3

1,2-Dicyanonaphthalene can also be prepared by a similar method from naphthalene-1,2-dicarboxylic anhydride, which is prepared from 1,2-dialkylnaphthalene or 1-chloromethyl-2-methylnaphthalene.[9] Kenney and co-workers prepared new types of bisalkylsiloxysilylnaphthalocyanines (4) in 1984.[10] 2,3-Dicyanonaphthalene is converted to 1,3-diiminobenz[f]-isoindoline, which tetramerizes to give dichlorosilylnaphthalocyanine (SiNcCl$_2$) in the presence of tetrachlorosilane. The acid hydrolysis of SiNcCl$_2$ give SiNc(OH)$_2$, which reacts with trialkylsilylchloride to give SiNc(OSiR$_3$)$_2$. The processes are shown in Scheme 4.

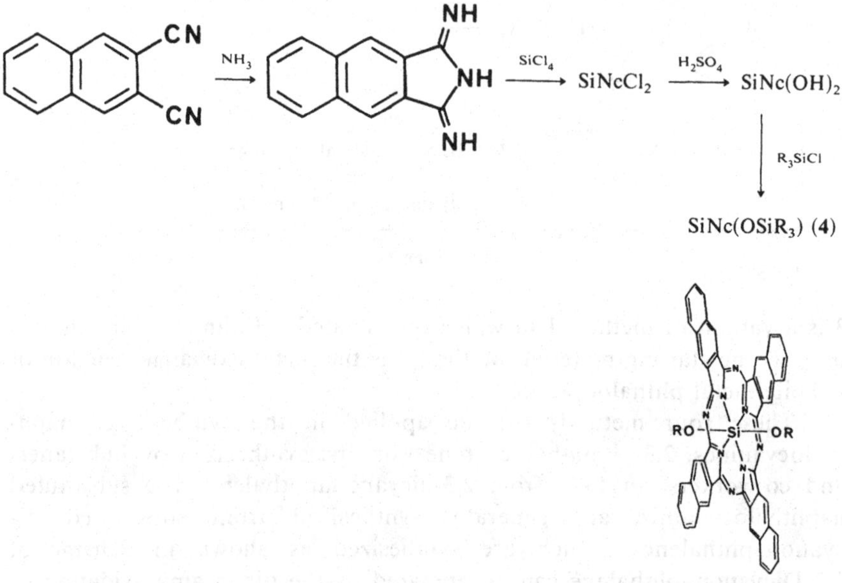

(4) R = SiR$_3$, COR, etc.

Scheme 4

Naphthalocyanines have two structural isomers: 1,2- (2) and 2,3-naphthalocyanines (3). Magnesium 1,2-naphthalocyanine was first reported by Bradbrook and Linstead[11] in 1936 and has two isomers, the α and β forms. Zinc and copper 1,2-naphthalocyanines have also been reported,[11] but their absorption spectra have not been described. Table I shows the elements of the periodic table for which phthalocyanine and 2,3-naphthalocyanine derivatives are available.

Table I. Elements Employed in the Preparation of Phthalocyanine (Pc)[a] and Naphthalocyanine (Nc)[b] Pigments[c]

In the table below, ◎ denotes an element with a double circle (phthalocyanine and naphthalocyanine derivatives available) and ○ denotes an element with a single circle (phthalocyanine derivative available).

Group	I	II	III	IV	V	VI	VII	VIII
1	H ◎							
2	Li ○	Be ○	B	C	N	O	F	
3	Na ○	Mg ◎	Al ◎	Si ○	P	S	Cl	
4	K ○, Cu ◎	Ca ○, Zn ○	Sc ○, Ga ○	Ti ◎, Ge ○	V ◎, As	Cr ◎, Se	Mn ◎, Br	Fe ◎ Co ◎ Ni ◎
5	Rb, Ag ○	Sr, Cd ○	Y ○, In ○	Zr ○, Sn ◎	Nb ◎, Sb ○	Mo ◎, Te	Tc ○, I	Ru ◎ Rh ◎ Pd ◎
6	Cs, Au ○	Ba ○, Hg ○	Tl ○	Hf ○, Pb ◎	Ta ○, Bi	W ○, Po	Re ○, At	Os ○ Ir ○ Pt ○
7	Fr	Ra						

Lanthanides: La ○ Ce ○ Pr ○ Nd ○ Pm ○ Sm ○ Eu ○ Gd ○ Tb ○ Dy ○ Ho ○ Er ○ Tm ○ Yb ○ Lu ○

Actinides: Th ○ Pa ○ U ○ Np ○ Am ○

[a] S. Tamura, Shikizai Kyokaishi 58, 528 (1985).
[b] Results from this work.
[c] ○, phthalocyanine derivative is available; ◎, phthalocyanine and naphthalocyanine derivatives are available.

III. COLOR AND CONSTITUTION

The structures of porphine and related compounds are summarized in Scheme 5.

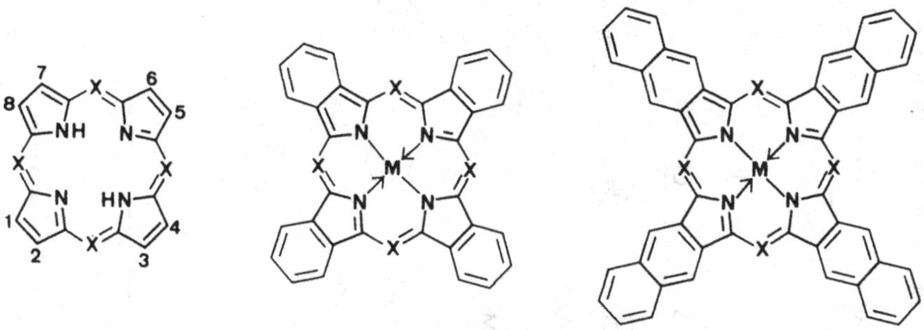

(5) X = CH
(6) X = N

(1) X = N
(7) X = CH

(3) X = N
(8) X = CH

Scheme 5

Phthalocyanine has the same chromophoric system as porphine (5), which is a planar, cyclic 16-center ring system consisting of four bridged pyrrole rings. Tetraazaporphine (6) is the nitrogen-substituted analogue, with nitrogen atoms at the four meso positions of 5. The outer four ethylene units are essentially pure double bonds, and they can be replaced by benzene rings to give tetrabenzoporphine (7). Substitution of the four meso carbons of 7 by nitrogen produces phthalocyanines (1), and another benzannelation of 1 and 7 gives naphthalocyanines (3) and tetra(2,3-naphtho)porphine (8), respectively. The 16-center ring system containing 18 π electrons is the basic chromophore of porphyrin. Benzannelation causes the bathochromic shift of the first absorption band, named the Q-band, from 619 nm (ε 4,570) for porphine to 698 (162,200) for phthalocyanine. The increase of the molar absorptivity in phthalocyanine is remarkable. The vibrational splitting of the Q-band resulted in two λ_{max} at 698 and 665 nm. Copper phthalocyanine shows λ_{max} at 678 nm with a molar absorptivity of 218,800, and a shoulder is observed at about 650 nm.[5]

The color–structure relationship of phthalocyanine chromophores has been evaluated by MO calculations.[5,12] Their λ_{max}, their molar absorptivity, and the π-electron density changes accompanying the first transition were calculated, and the calculations indicated that the π electrons migrate from the center toward the outside. Metalation, which reduces the electron density at the inner nitrogen atoms, is predicted to produce a hypsochromic shift, and this is observed experimentally. The extent of the λ_{max} shift to shorter

Table II. Absorption Spectra of Porphine (Pp), Phthalocyanine (Pc), and Naphthalocyanine (Nc) Derivatives

Compound	λ_{max} (ε_{max}) (nm)	Ref.
Porphine (**5**)	619 (4,570), 529 (7,586)	a
Tetraazaporphine (**6**)	624 (81,300), 556 (46,770)	a
Tetrabenzoporphine (**7**)	662 (15,800), 596 (21,800)	b
Phthalocyanine (**1**)	698 (162,200), 665 (151,400)	a
Tetra(2,3-naphtho)porphine (**8**)	722, 703	c
2,3-Naphthalocyanine (**3**)	765 (182,000)	
t-Octyl-VO-naphthalocyanine	809 (245,000)	d
1,2-Naphthalocyanine (**2**)	720, 677	e

[a] Ref. 5, p. 223.
[b] M. Hanack and T. Zipplies, *J. Am. Chem. Soc.* **107**, 6127 (1985).
[c] M. Rein and M. Hanack, *Chem. Ber.* **121**, 1601 (1988).
[d] M. Matsuoka, *Absorption Spectra of Dyes for Diode Lasers*, Bunshin, Tokyo (1990).
[e] J. S. Anderson, E. F. Bradbrook, A. H. Cook, and R. P. Linstead, *J. Chem. Soc.*, 1151 (1938).

wavelength depends on the electronegativity of the metal. In contrast, electron-withdrawing groups, such as halogens, at the periphery of the molecule are predicted to produce a bathochromic shift. Annelation also produces a bathochromic shift. However, in principle, the 18-π-electron inner ring system of the phthalocyanines makes the main contribution to bathochromicity, and the substitution in the benzene rings has only a minor effect on the λ_{max} of phthalocyanines.

Tetraphenylphthalocyanine absorbs at 715 nm, and a bathochromic shift of about 20 nm is produced by the introduction of a phenyl group into the benzene ring. Benzannelation of phthalocyanine produces quite a large bathochromic shift of 67 nm in **3** and 22 nm in **2** and increases the molar absorptivity due to the enlargement of the π-conjugating systems. A comparison of the absorption spectra of various compounds in the porphine series is presented in Table II. The effects of various metals on the absorption spectra of phthalocyanine and related compounds are summarized in Table III.

IV. APPLICATIONS OF PHTHALOCYANINES AND NAPHTHALOCYANINES

A. Applications of Phthalocyanines

Phthalocyanine (Pc) compounds have drawn attention as functional dyes in various fields, and much interesting research has been reported.[13]

Table III. Absorption maxima of Metal Phthalocyanines and Related
Compounds

Metal	λ_{max} (nm)				
	1[a]	2	3	7[b]	8
H	686	720	765	662	722
Zn	661				699
Co	657				689
Al(F)			717[c]		
Co(CN)					702
Co(Cl)					696
Fe	676	640[d]			
Mg	666	685	720[c]	628	
Cd				628	
Cu	658				
Pb	698				

[a] L. Edwards and M. Gouterman, *J. Mol. Spectrosc. 33*, 292 (1970).
[b] R. B. M. Koehorst, J. F. Kleibeuker, T. J. Schaafsma, D. A. de Bie, B. Geurtsen, R. N. Henrie, and H. C. van der Plas, *J. C. S.* 1005 (1981).
[c] Y. Shimura, M. Hoshi, and M. Shimura, *J. Electrochem. Soc.* 239 (1986) (values for vacuum-deposited film).
[d] G. M. Magner, M. Savy, and G. Scarbeck, *J. Electrochem. Soc. 127*, 1076 (1980).

Their semiconductor or dielectric properties are useful for diode, piezo-electric, and electrochromic devices. Their optical properties make them suitable for practical applications in xerography, photochemical hole burn-ing, laser disk memory, and photodiodes. Their catalytic activities have led to their utilization in photovoltaic cells and chemical sensors.

The semiconductivity of phthalocyanines was first reported indepen-dently by Eley[14] and Vartanyan[15] in 1948. In 1960, Brenner[16] mentioned that probably the earliest studies of organic substances as intrinsic semicon-ductors were started in 1948 when Eley and Vartanyan discovered the unusual temperature dependence of the resistivity of phthalocyanines and their metal derivatives. Their photoelectric sensitivity was investigated by Putseiko[17] in 1949. Since then, many dye chromophores have been reported as organic photoconductors (OPC) which show a good sensitivity over wide ranges of visible wavelengths.

Phthalocyanines are now widely applied as OPC for electrophotogra-phy (xerography) and laser printer systems (Chapter 12). The conductivity and excitation energies of phthalocyanine and its derivatives are summarized in Table IV.

Table IV. Resistivity (ρ) and Excitation Energy of
Phthalocyanines (Pc)

MPc	$\rho(\Omega\,cm)$	E (eV)	E'(eV)	Ref.
H_2Pc	3.0×10^{13}	1.82	0.50	a, b
$CuPc(\beta)$	1.2×10^{13}	1.70	0.42	a, c
$CuPc(\varepsilon)$	—	0.98	0.96	c
MgPc	1×10^6	—	—	d, e
FePc	4×10^9	0.72	0.40	a, f
MnPc	4×10^6	—	—	c, f
$NiPc(\beta)$	6.3×10^{10}	1.92	0.82	c, f
$ZnPc(\beta)$	2.8×10^9	1.96	0.64	c, f
VOPc	5.0×10^{13}	1.48	0.94	c

[a] A. Braun and J. Tcherniac, *Chem. Ber. 40, 2709 (1907).*
[b] G. H. Heilmeier and S. E. Harrison, *Phys. Rev. 132,* 2010 (1963).
[c] S. Tamura, *Shikizai Kyokaishi 58,* 528 (1985).
[d] B. D. Linstead, *J. Chem. Soc.,* 1719 (1936).
[e] E. Putseiko, *Dokl. Akad. Nauk. SSSR 59,* 471 (1948).
[d] J. Yamashita and T. Kurosawa, *J. Phys. Soc. Jpn. 15,* 802 (1960).

B. Applications of Naphthalocyanines

Naphthalocyanines (Nc) absorb in the near-infrared region and are interesting as recording media for laser optical recording systems. The most important question is how to get good solubility in nonpolar organic solvents. Dye media are dissolved in an organic solvent in high concentration and then applied onto the substrate made of polycarbonate or other polymers to produce the recording disk. In the first stage of development, good solubility was achieved by the introduction of a long alkyl chain into the naphthalene ring. The synthesis of these derivatives is shown in Scheme 3. Solubility in organic solvents is increased by preventing the aggregation of Nc caused by the steric interaction between the bulky and long-chain alkyl substituents. Some examples of substituted Nc derivatives are shown in Scheme 3. They are easily dissolved in nonpolar organic solvents such as methyl cellosolve.

In 1984, Kenney and co-workers[10] synthesized the silicon naphthalocyanines, which opened the new field of Nc chemistry. The synthetic routes for solubilized Nc are shown in Scheme 4. The solubility can be controlled by the alkylsilyl substituents. Some metal phthalocyanines such as Fe(III), Al(III), and cobalt phthalocyanine have the substituent or other ligands outside the Pc π-conjugated plane. In the case of silicon derivatives, another two bonding electrons are coordinated above and below the π-conjugated plane and are substituted with bulky substituents such as trialkylsiloxyl groups. Each of the silicon naphthalocyanines cannot be aggregated to each other because of the steric hindrance of the alkylsilyl groups and thus becomes soluble in high concentration in organic solvents. Typical examples of silicon Nc are shown in Scheme 4. Their thin films absorb at 750–830 nm and have a reflectance of 30% at 800 nm. These characteristics of Nc thin films are favorable for optical recording media in practical uses. Many Nc derivatives have been patented as infrared absorbing dyes for optical recording media, organic photoconductors, and sensors. The number of Japanese patents on Nc has been increasing steadily in the past few years, from 2 in 1985 to 7 in 1986, 9 in 1987, and 14 in 1988.

Silicon naphthalocyanines have been developed because of interest in their use as photodynamic sensitizers. Some metal phthalocyanines and their naphthalocyanine derivatives have been investigated as photodynamic sensitizers to produce hydrogen by electrolysis of water, thus providing a clean energy source to replace hydraulic or nuclear sources. Their redox behavior, reversible energy-transfer reactions, and electrical conductivity has been investigated in relation to their structures and spectral properties.

Savy and co-workers[18] recently reviewed the problem of water electrolysis in which a new type of molybdenum 1,2-naphthalocyanine played an effective catalytic role. 1,2-MoNc has better catalytic activity than 1,2-FeNc in the oxygen evolution by water electrolysis experiment. They prepared metal 1,2-naphthalocyanines and investigated their structures by X-ray photoelectron studies in relation to the electrocatalysis of oxygen reduction and evolution.[19] Their interests are related to the catalytic roles of the porphyrin series in biological systems and also for their efficiency as catalysts for air electrodes.

Kenney and co-workers[20] synthesized silicon naphthalocyanines and evaluated their utility as photodynamic sensitizers for photodynamic therapy of tumors (Chapter 15). Singlet molecular oxygen, generated by energy transfer from the triplet state of dye chromophores such as silicon naphthalocyanines to ground-state triplet molecular oxygen, is regarded as a leading candidate for the initiation of tissue damage in the presence of light, oxygen, and an absorber. Kenney and co-workers studied the electrochemistry, electrogenerated chemiluminescence, and a reversible energy-transfer reaction of phthalocyanines and naphthalocyanines.[10] They have been interested in the conductivities of these materials[21] and in their application to the sensitization of semiconductor electrodes in photoelectrochemical cells.[22] They have been interested in the synthesis of group IV (silicon) phthalocyanines and naphthalocyanines because various groups can be attached to the axial positions. Further, some group IV phthalocyanines form linear stacked polymers.

REFERENCES

1. R. P. Linstead, *J. Chem. Soc.*, 1016, 1017, 1022, 1027, 1031, 1033 (1934).
2. J. M. Robertson, *J. Chem. Soc.*, 615 (1935); 1195, 1736 (1936); 219 (1937); 36 (1940).
3. F. H. Moser and A. L. Thomas, *Phthalocyanine Compounds*, Reinhold, New York (1963).
4. G. Booth, in: *The Chemistry of Synthetic Dyes* (K. Venkataraman, ed), Vol. 5, p. 241, Academic Press, New York (1971).
5. P. F. Gordon and P. Gregory, in: *Organic Chemistry in Colour*, p. 219, Springer-Verlag, London (1983).
6. L. Edwards and M. Gouterman, *J. Mol. Spectrosc. 33*, 292 (1970).
7. E. I. Kovshev, V. A. Puchnova, and E. A. Luk'yanets, *Zh. Org. Khim. 7*, 369 (1971).
8. Nihon Jyoryu Kogyo Co., Ltd.
9. M. Matsuoka, M. Kitano, T. Kitao, and K. Konishi, *Nippon Kagaku Kaishi*, 1323 (1973).
10. B. L. Wheeler, G. Nagasubramanian, A. J. Bard, L. A. Schechtman, D. R. Dininny, and M. E. Kenney, *J. Am. Chem. Soc. 106*, 7404 (1984).
11. E. F. Bradbrook and R. P. Linstead, *J. Chem. Soc.*, 1744 (1936).
12. J. Griffiths, *Colour and Constitution of Organic Molecules*, p. 232, Academic Press, London (1976).
13. S. Tamura, *Shikizai Kyokaishi 58*, 528 (1985).
14. D. D. Eley, *Nature 162*, 819 (1948).
15. A. T. Vartanyan, *Zh. Fiz. Khim. 22*, 769 (1948).
16. W. Brenner, *Mater. Design Eng. 51*, 12 (1960).
17. E. K. Putseiko, *Dokl. Akad. Nauk. SSSR 67*, 1009 (1949).
18. M. Dieng, O. Contamin, and M. Savy, *Electrochim. Acta 33*, 121 (1988).
19. J. Riga, M. Savy, J. J. Vervist, J. E. Guerchais, and J. Salapala, *J. C. S. Faraday Trans. 1, 78*, 2773 (1982).
20. P. A. Firey, W. E. Ford, J. R. Sounik, M. E. Kenney, and M. A. J. Rodgers, *J. Am. Chem. Soc. 110*, 7626 (1988).
21. R. S. Nohr, P. M. Kuznesof, K. J. Wynne, M. E. Kenney, and P. G. Siebenman, *J. Am. Chem. Soc. 103*, 4371 (1981).
22. P. Leempoel, M. Castro-Acuna, F. R. F. Fan, and A. J. Bard, *J. Phys. Chem. 86*, 1396 (1982).

6

Metal Complex Dyes

KENRYO NAMBA

I. INTRODUCTION

Many kinds of metal complexes have been synthesized for the chemical analysis of metal ions and for use as catalysts for organic reactions, as dyes and pigments, and in other materials. Many metal complexes are also used as mordant dyes. They are, in general, insoluble or slightly soluble in water and most organic solvents. Some kinds of nickel complexes have been used for the Q-switch dye laser,[1] in infrared absorbing films for agricultural uses, and as infrared absorbers in sunglasses and goggles, antioxidants for polymers,[2,3] and singlet oxygen quenchers for the photofading of dyes.[4]

Recently developed semiconductor laser diodes, which emit laser light at 760–850 nm, requires new near-infrared absorbing dyes as light absorbers. The optical memory disk is the most interesting application of these dyes. Many metal complex dyes have been prepared as dye media or singlet oxygen quenchers for optical memory disks.

This chapter describes several types of metal complexes used as near-infrared absorbing dyes.

KENRYO NAMBA • Advanced Materials Research Department, R & D Center, TDK Corporation, Ichikawa, Chiba 272, Japan.

II. DITHIOLENE METAL COMPLEXES

A. Dithiobenzil and Ethylenedithiolate Metal Complexes

Schrauzer first synthesized dithiolene metal complexes under mild conditions applying Steinkopf's historical thiophene synthesis of 1,2-diphenylacetylene with NiS_x.[5] The proposed structure 1 was determined by X-ray analysis.[6] Many transition metal complexes have been prepared by treatment of acyloins, especially benzoin derivatives, with P_4S_{10}.[7] Equation (1) shows the method for the synthesis of bisdithiobenzil metal complexes.

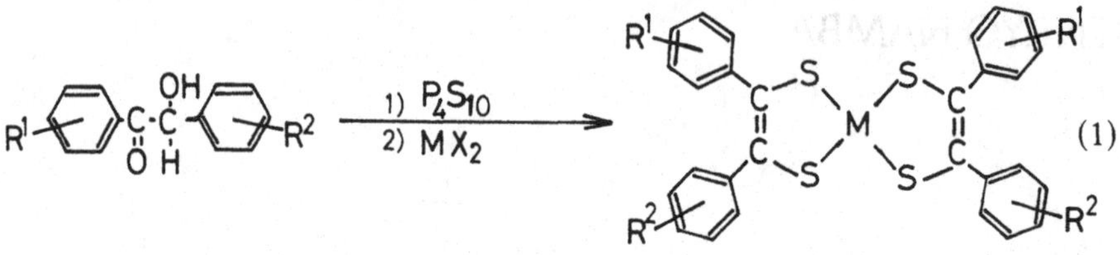

$$\text{(1)}$$

(1)

Metal complex 1 was also obtained from the reaction of sulfur with diphenylacetylene and nickel or nickel tetracarbonate.[8] The yields are generally low but are improved by the addition of ammonium sulfate.[9] Another improved method, shown in Scheme 1, has been reported for the preparation of mesomorphic transition metal complexes having *para* alkyl substituents.[10]

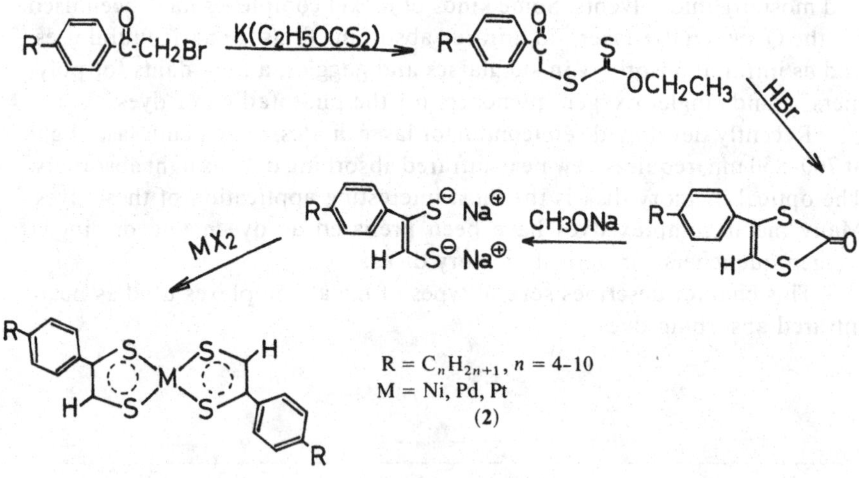

$R = C_nH_{2n+1}$, $n = 4$–10
$M = $ Ni, Pd, Pt

$$\text{(2)}$$

Scheme 1

Table I. Infrared Absorption Bands of Dithiolene Metal Complexes

$$R^1 \underset{R^2}{\overset{S}{\diagdown}} \underset{S}{\overset{S}{\diagup}} \overset{R^2}{\underset{R^2}{M}}$$

R^1	R^2	Metal	λ_{max} (nm)	log ε	Solvent	Solubility in benzene at 25°C (wt %)	Ref.
CF_3	CF_3	Ni	715	(4.09)	n-C_5H_{12}		11
C_6H_5	C_6H_5	Pd	885	(4.61)	$CHCl_3$		11
C_6H_5	C_6H_5	Pt	802	(4.63)	$CHCl_3$		11
C_nH_{2n+1} (n = 4–10)	H	Ni	850	—	n-C_6H_{14}		10
C_nH_{2n+1} (n = 4–10)	H	Pd	865	—	n-C_6H_{14}		10
C_nH_{2n+1} (n = 4–10)	H	Pt	780	—	n-C_6H_{14}		10
C_6H_5	C_6H_5	Ni	855	4.48	CH_2Cl_2	0.13	14
p-$CH_3C_6H_4$	p-$CH_3C_6H_4$	Ni	877	4.50	CH_2Cl_2	0.15	14
p-$OCH_3C_6H_4$	p-$OCH_3C_6H_4$	Ni	894	4.45	CH_2Cl_2	0.12	14
p-ClC_6H_4	p-ClC_6H_4	Ni	861	4.55	CH_2Cl_2	0.12	14
p-$CF_3C_6H_4$	p-$CF_3C_6H_4$	Ni	832	4.48	CH_2Cl_2	0.82	14
3,4-$Cl_2C_6H_3$	3,4-$Cl_2C_6H_3$	Ni	850	4.49	CH_2Cl_2	0.82	14
o-ClC_6H_4	o-ClC_6H_4	Ni	783	4.41	CH_2Cl_2	0.56	14
o-BrC_6H_4	o-BrC_6H_4	Ni	783	4.40	CH_2Cl_2	1.49	14
2,4-$Cl_2C_6H_3$	2,4-$Cl_2C_6H_3$	Ni	787	4.35	CH_2Cl_2	4.50	
CH_3	CH_3	Ni	771	4.32	CH_2Cl_2	0.13	14
2-Thienyl	2-Thienyl	Ni	995		DMF		17
p-$N(CH_3)_2C_6H_4$	C_6H_5	Ni	1125		DMF		17
p-$N(CH_3)_2C_6H_4$	p-$NH_2C_6H_4$	Ni	1163		DMF		17

The formation of these dithiolene metal complexes involves the interaction between the lowest unoccupied π molecular orbitals of the ligand and the occupied orbitals of the central metal. The light absorption in the near-infrared region is assigned to the first allowed π–π transition between $2b_{1u}$ and $2b_{2g}$ orbitals and involves molecular orbitals extending over the whole molecule, as indicated by its high intensity and large substituent effects.[11,12] The assignment of the absorption band is based on results from resonance Raman spectroscopy[13] and evaluation of absorption spectra by the Pariser–Parr–Pople (PPP) MO method.[12,13]

Table I shows the absorption maxima of several dithiolene metal complexes. Bis(1,2-diaryl-1,2-ethylenedithiolato) metal complexes (1) absorb at longer wavelengths than bis(1,2-dialkyl-1,2-ethylene-dithiolato) metal complexes. The introduction of halogen atoms at the 2-position[14,15] or of dialkylamino groups at the 4-position[16] in the benzene rings of diaryl metal complexes produced a hypsochromic shift and a bathochromic shift, respectively, of the near-infrared absorption band. The hypsochromic shift is caused by the steric hindrance between the benzene rings and the chelate ring, which decreases the molar absorptivity. Molecular design using this type of steric hindrance is very important in adapting the λ_{max} for the laser diode.

The solubility of the metal complexes in organic solvents, which is very important for the spin coating process in the production of optical disks, was also improved by the introduction of long alkyl groups such as an octyl group at the 4-position[16] or of a halogeno or methyl group at the 2-position. Solvent effects on λ_{max} of these complexes were studied by Freyer[17,18] and Graczyk.[19] Graczyk explained the spectral changes in terms of a change in the molecular symmetry of dyes such as the bis(4-dimethyl-aminodithiobenzil)nickel complex [1; R = NH(CH$_3$)$_2$]. The molecular symmetry changes from square planar to rhombic bipyramidal due to a change in the electronic configuration of the central nickel ion from $d_{x^2}^2$ to $d_{z^2}^1$, $d_{x^2-y^2}^1$.

Table II shows the effect of solvents on the infrared absorption bands of the bis(4-dimethylaminodithiobenzil)nickel complex.[19]

Bis-camphenyldithio metal complexes which absorb at 768–837 nm, the most favorable wavelength region for laser diodes, and have good solubility in organic solvents such as toluene and dichloromethane have been reported.[20]

Table II. Absorption Bands in Near-Infrared Region of Bis(4-dimethylaminodithiobenzil)nickel Complex in Various Solvents[a]

Solvent	DN[b]	λ (nm)	log ε	λ (nm)	log ε
CH$_2$Cl$_2$	0.1	1072	4.03		
CH$_3$NO$_2$	2.7	1080	4.28	1140	4.29
C$_6$H$_5$NO$_2$	4.4	1088	4.43	1124	4.47
C$_6$H$_5$Cl	9.0	1068	4.69	1128	4.64
CH$_3$CN	14.1	1075	4.47		
Dioxane	14.8			1152	2.48
THF	20.0	1058	4.36		
DMF	27.0	1040	3.56	1172	3.33
DMSO	29.8	1096	4.49		

[a] Ref. 19.
[b] DN: donor number of the solvent.

B. Benzene-1,2-dithiolate Metal Complexes

Benzene-1,2-dithiol derivatives easily react with metal salts to give square-planar bis(benzene-1,2-dithiolate) metal complexes **3** (M = Ni, Cu, Co, Pt; R = Cl, Br, Me, NMe$_2$).[21,22]

(3)

Various ligands such as toluene-3,4-dithiolate,[21] benzene-1,2-dithiolate, o-xylene-4,5-dithiolate, 3,4,5,6-tetramethylbenzene-1,2-dithiolate, prehnitene-5,6-dithiolate, and 3,4,5,6-tetrachlorobenzene-1,2-dithiolate[22] were reacted with some metal ions including cobalt, nickel, and copper, and their magnetic, polarographic, and spectral properties investigated. 3,4,5,6-Tetrachlorobenzene-1,2-dithiol was synthesized by the reaction of hexachlorobenzene with sodium sulfhydrate in the presence of iron powder.[22] Mono-, di-, and trihalogenated benzene-1,2-dithiol ligands can be prepared with improved yields in the presence of sulfur.[23] 4-Dimethylaminobenzene-1,2-dithiolate was also synthesized as a new ligand to act as a near-infrared absorber and singlet oxygen quencher.[24] These ligands gave square-planar metal complexes which have better solubility in polymer or organic solvents. Only the nickel complexes of benzene-1,2-dithiol derivatives absorb in the near-infrared region, as shown in Table III. In general, they absorb at much longer wavelengths than the oscillating wavelength of the laser diode, and, except for some complexes such as the bis(3,4,6-trichlorobenzene-1,2-dithiolate)nickel complex, their solubility in organic solvents is not sufficient for spin coating applications.

III. PHENYLENEDIAMINE METAL COMPLEXES

The phenylenediamine metal complexes **4** (M = Ni; R^1 = Me, Et, Bu, NH$_2$, NO$_2$; R^2 = H, Et) exhibit large bathochromic shifts of over 60 nm

(4)

Table III. Spectral Properties of Bis(benzenedithiol) Metal Complexes

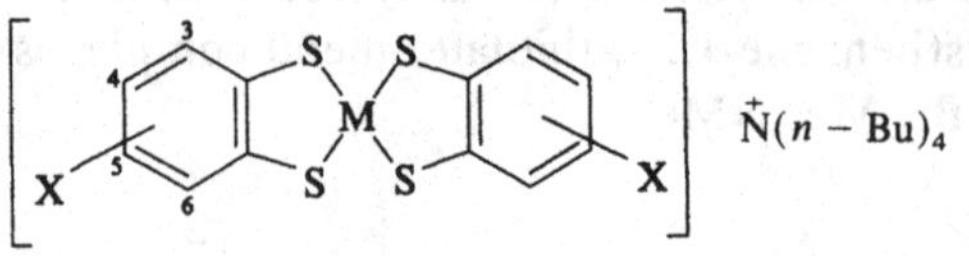

X	Metal	λ_{max} (nm)	Solvent[a]	log ε	Ref.
Me$_4$	Ni	926	A	4.21	22
4,5-Me$_2$	Ni	917	A	4.19	22
4-Me	Ni	890	B	—	21
Cl$_4$	Ni	885	A	4.20	22
H	Ni	881	A	4.12	22
4-Me	Co	658	B	—	21
4-Me	Cu	641	B	—	21
4-Me	Fe	562[b]	C	—	21

[a] A: CH_2Cl_2, CH_3CN, pyridine, acetone, DMF, or DMSO; B: CH_2Cl_2, CH_3CN, or pyridine; C: CH_2Cl_2 or acetone.
[b] Solvatochromism was observed as follows: DMF, 498; DMSO, 491; pyridine, 515; CH_3CN, 525 nm.

compared with the dithiolene metal complexes. Introduction of an ethyl group into the benzene ring of the bisphenylenediamine nickel complex improved solubility in organic solvents, and the complex with amino and butyl substituents is soluble in alcohol and absorbs at 790–795 nm, the most favorable wavelength for laser optical memory material. Light stability is the same as or higher than that of dithiolene complexes.[26]

Mixed-ligand metal complexes containing imine, selenol, and thiol groups as shown in Scheme 2 have been synthesized and their spectral properties reported.[27]

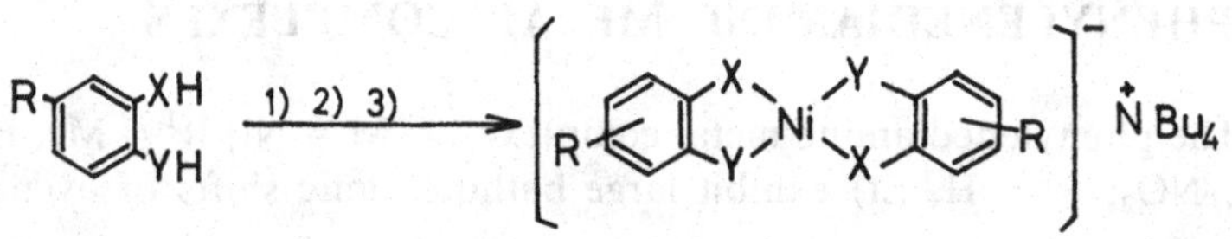

(5)

(1) KOH or K; (2) $NiCl_2 \cdot 6H_2O$; (3) $(n\text{-}C_4H_9)_4NBr$
X = S, NH;   Y = S, Se

Scheme 2

IV. INDOANILINE-TYPE METAL COMPLEXES

Metal complexes synthesized from N,O-bidentate indoaniline-type ligands absorb near-infrared light at 745–776 nm in ethanol.[28] The ligands were synthesized by the condensation reaction of 8-hydroxyquinoline and dialkylaminoaniline hydrochlorides (Scheme 3). They easily formed com-

(6)

M = Ni, Cu

Scheme 3

plexes having a quinonoid structure with metal salts in solution under nitrogen atmosphere. Absorption maxima of the complexes are listed in Table IV. The molar absorptivities of these new complexes are remarkably high, about 10 times those of dithiolate metal complexes. This may contribute to the high reflectance of thin films of dye layer in the near-infrared region (50% at 900 nm), as shown in Figure 1.[28]

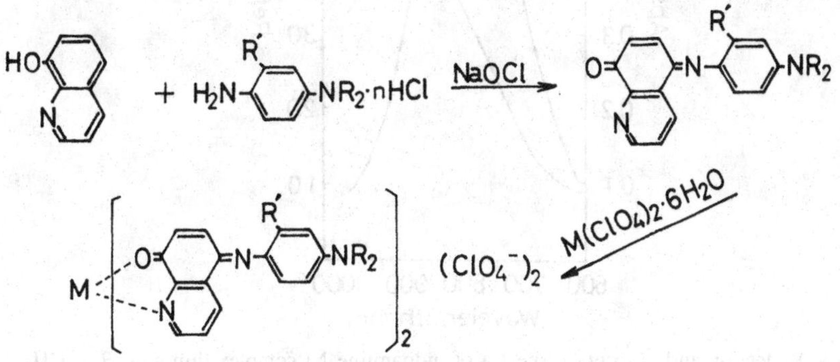

Table IV. Spectral Data for Indoaniline Metal Complexes 6^a

R	R′	Metal	λ_{max} (nm)^b	$\log \varepsilon$
Et	Me	Cu	776	5.16
Et	Me	Ni	775	5.05
Me	H	Cu	772	5.16
Me	H	Ni	745	4.93

^{a} Ref. 28.
^{b} Solvent: EtOH.

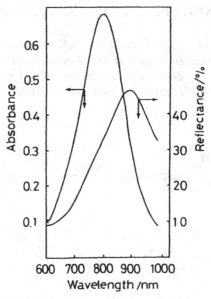

Figure 1. Reflection and absorption spectra of indoaniline Ni complex thin film; $R' = CH_3$, $R = C_2H_5$, thickness = 50 nm.

V. APPLICATIONS OF INFRARED ABSORBING METAL COMPLEX DYES

The most interesting application of near-infrared absorbing dyes is their use as light absorbing media in laser optical memory systems. As shown in Figure 2, the initial composition of an optical memory disk was double layered[29]: dye was used for the light absorbing layer and metal for the reflective layer, which is essential to reproduce data by the optical method (cf. Chapter 10). The metal reflection layer causes problems such as difficulty in reading and writing through a substrate, low sensitivity in recording because of the high thermal conductivity of the metal layer, and difficulties in the production process. The thin-layer film prepared from some kinds of cyanine dye reflects near-infrared light very well (over 50%), and it is possible to produce a monolayer disk which can be written on and read through the substrate. However, thin films composed of metal complexes such as dithiobenzil and benzenedithiol metal complexes have low reflectivity (at most 15%).

High reflectivity and strong absorption of near-infrared light can be achieved with thin films made of indoaniline metal complex and applied to the monolayered optical disk.

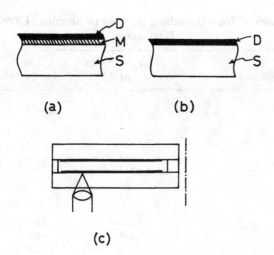

Figure 2. Structure of recording layer of dye optical disk: (a) double layer of dye and metal; (b) monolayer of dye; (c) protective air sandwich structure of optical disk. D: Recording layer of dye; M: reflective layer of metal; S: substrate.

Other significant applications of these metal complexes are their use as singlet oxygen quenchers to prevent the photodegradation of polymers and the photofading of dyes. Many types of metal complexes have been investigated as singlet oxygen quenchers by Foote.[30] The nickel complexes are the most effective quenchers, as shown in Table V, but their use is very restricted because of their color. They absorb a wide range of light from the visible to the near-infrared region and color substrates such as polyolefin and photographic materials. Many new dithiolene nickel complexes that are readily soluble in organic solvents for spin coating have been synthesized as singlet oxygen quenchers for cyanine dye thin films in optical disks, for which visible light absorption of the quencher is not a problem. Dithiobenzil and benzenedithiol nickel complexes which absorb near-infrared light and are soluble in organic solvents are more effective as quenchers than any other complex. The quenching effect of the dithiolene nickel complexes has been investigated by Nakazumi et al.[31] in cyanine dye thin films. They have reported that bis(trimethoxyphenyldithiobenzil)nickel is the most soluble and effective singlet oxygen quencher for infrared absorbing cyanine dyes in thin films for optical disks. The effect of substituents on the lightfastness of three types of cyanine dyes in methylene chloride solution has also been investigated in the absence and presence of bis(4-dimethylamino-dithiobenzil)nickel. The lightfastness of any type of cyanine dye is improved in the presence of the nickel complex, as shown in Figure 3.[32]

Table V. Rates of Total Quenching ($k_q + k_r$) or Reaction (k_r) for Various Compounds

Compound	$(k_q + k_r) \times 10^{-6}$ $(M^{-1}\,s^{-1})$	$k_r \times 10^{-6}$ $(M^{-1}\,s^{-1})$	Solvent
Carotenes			
β-Carotene	1.2×10^4		C_6H_6
Amines			
$(C_2H_5)_2NH$	1.9		CH_3OH
	15		$CHCl_3$
DABCO	17		CH_3OH
	52		$CHCl_3$
	40		C_6H_6
Phenols			
α-Tocopherol	530	36	CH_3OH
	148	~1.5	C_6H_6
	250	2.1	Pyridine
2,4,6-Triphenylphenol			
	239	very small	CH_3OH
	21.3		C_6H_6
Ni complexes			
Dithiobenzil	2.2×10^4		Toluene
Bis(diisopropyldiethiophosphato)	7.6×10^3		CCl_4

The quenching effect has been investigated quantitatively with the parameters $k_r + k_q$ (Table V) in the oxidation reaction of rubrene (Ru):

$$Ru \xrightarrow{h\nu} Ru^1 \tag{2}$$

$$Ru^1 + O_2 \xrightarrow{k_s} Ru^3 + {}^1O_2 \tag{3}$$

$$Ru^3 + O_2 \xrightarrow{k_t} Ru + {}^1O_2 \tag{4}$$

$${}^1O_2 + Ru \xrightarrow{k_o} RuO_2 \tag{5}$$

$${}^1O_2 \xrightarrow{k_d} {}^3O_2 \tag{6}$$

$${}^1O_2 + Q \xrightarrow{k_q} {}^3O_2 + Q \tag{7}$$

$${}^1O_2 + Q \xrightarrow{k_r} QO_2 \tag{8}$$

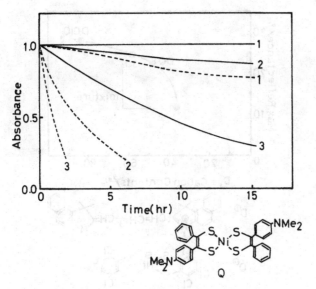

Figure 3. Effect of substituents on the lightfastness of croconium (1), squarylium (2), and cyanine dyes (3) in solution. Irradiated by 1.2-kW xenon lamp. ————, Dye (1×10^{-4} mol litre^{-1}) in CH$_2$Cl$_2$; ———, Dye with nickel complex Q (2:1) in CH$_2$Cl$_2$.

		P	X	Y
1	Croconium dye	$-CH=$ (croconium ring)	Se	H
2	Squarylium dye	$-CH=$ (squarylium ring)	CH=CH	H
3	Cyanine dye	$(CH=CH)_2$	CH=CH	H

where Q represents the quencher. Nakazumi *et al.* obtained k_r and k_q and showed that k_q remains almost constant but that k_r is affected by the substituents of bis(dithiobenzil)nickel complexes.

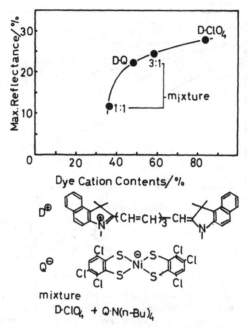

Figure 4. Effect of dye cation content on the maximum reflectance of thin films.

Bis(benzenedithiol) derivatives of nickel complexes have also been reported to be superior quenchers for dye media in the optical disk. However, the amount of quencher which can be added to the cyanine dye film is limited, because the reflectivity of the thin film is severely decreased as the amount of quencher added is increased, as shown in Figure 4. The reflectivity of a thin film composed of equimolecular mixtures of cyanine dye and benzenedithiol nickel complex decreased to about one-third of that of cyanine dye. Namba has reported a highly reflective, light-resistant cyanine dye film for optical disks consisting of the ionic salt formed between the cationic cyanine dye and the benzeneditiol nickel complex anion (7). This

(7)

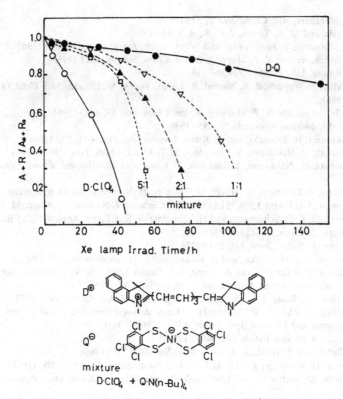

Figure 5. Effect of formation of a salt between cyanine cation (D⁺) and quencher anion (Q⁻) on lightfastness of thin films. A_0 = Absorbance before irradiation; R_0 = reflectance before irradiation; A = absorbance after irradiation; R = reflectance after irradiation.

type of cyanine dye is very stable against photon mode degradation, both by environmental light (Figure 5) and by the reading light of the laser diode, compared with the conventional cyanine dyes with, for example, perchlorate and halogen anions.[33]

REFERENCES

1. K. H. Drexhage and U. T. Muller-Westerhoff, *IEEE J. Quantum Electronics QE-8*, 759 (1972); K. H. Drehage and G. A. Reynolds, *Opt. Commun. 10* (1), 18 (1974).
2. A. Zweig and W. A. Henderson, *J. Polym. Sci. Polym. Chem Ed. 13*, 717 (1975).
3. S. M. Bloom, U.S. Patent 3,588,216 (1969).
4. K. Namba, Eur. Patent Appl. 0147083 B1 (1988); T. Satoh *et al.*, U.S. Patent 4,626,496 (1986).

5. G. N. Schrauzer, *Acc. Chem. Res. 2*, 72 (1969).
6. D. Sartain and M. R. Truter, *J. C. S., A*, 1264 (1967).
7. G. N. Schrauzer, V. P. Mayweg, and W. Heinrich, *Inorg. Chem. 4*, 1615 (1965).
8. G. N. Schrauzer and V. P. Mayweg, *J. Am. Chem. Soc. 84*, 3221 (1962).
9. S. M. Bloom, U.S. Patent 3,894,069 (1975).
10. U. T. Mueller-Westerhoff, A. Nazzal, R. J. Cox, and A. M. Giroud, *Mol. Cryst.* (*Letters*), 249 (1980).
11. G. N. Schrauzer and V. P. Mayweg, *J. Am. Chem. Soc. 87*, 1483 (1965).
12. J. Griffiths, *Shikizai Kyokaishi 59*, 485 (1986).
13. H. Nakazumi, H. Shiozaki, and T. Kitao, *Spectrochim. Acta 44A*, 209 (1988).
14. H. Shiozaki, H. Nakazumi, Y. Nakado, and T. Kitao, *Chem. Lett.*, 2393 (1987).
15. H. Shiozaki, H. Nakazumi, Y. Nakado, T. Kitao, and M. Ohizumi, *Chem. Express* , 61 (1988).
16. W. Schrott, P. Newman, B. Albert, M. Thomas, H. Barzynski, K. D. Schomann, and H. Kippelmoie, Ger Patent 3,508,751A1 (1986); W. Schrott, P. Newman, B. Albert, M. Thomas, H. Barzynski, K. D. Schomann, and H. Kippelmaire, Eur. Patent Appl. 0192215 B1 (1986).
17. W. Freyer, *Z. Chem. 24*, 32 (1984).
18. W. Freyer, *J. Prakt. Chem. 328*, 253 (1986).
19. A. Graczyk, E. Biakkowska, and A. Konarzewski, *Tetrahedron 38*, 2715 (1982).
20. W. Schrott, P. Neumann, and B. Albert, Ger. Patent 3,505, 750 A1 (1986); W. Schrott, P. Neumann and B. Albert, Eur. Patent Appl. 0193774 A1 (1986).
21. R. Williams, E. Billig, J. H. Waters, and H. B. Gray, *J. Am. Chem. Soc. 88*, 43 (1966).
22. M. J. Baker-Hawkes, E. Billig, and H. B. Gray, *J. Am. Chem. Soc. 88*, 4870 (1966).
23. K. Sasagawa and M. Imai, Jpn. Patent A58-105960 (1983).
24. I. Tabushi *et al.*, Jpn. Patent B62-44036 (1987).
25. A. L. Balch and R. H. Holn, *J. Am. Chem. Soc. 88*, 5201 (1966).
26. A. Vogler, H. Kumkely, J. Hlavatsch, and A. Merz, *Inorg. Chem. 23*, 506 (1984).
27. S. H. Kim, M. Matsuoka, M. Yomoto, Y. Tsuchiya, and T. Kitao, *Dyes Pigments 8*, 381 (1987).
28. Y. Kubo, K. Sasaki, and K. Yoshida, *Chem. Lett.*, 1563 (1987).
29. A. Bloom, W. Joseph Burke, and D. Louis Ross, Jpn. Patent A56-92095 (1981); K. Sasagawa, E. Noda, and M. Imai, Jpn. Patent 1299013 (1985).
30. C. S. Foote, in: *Singlet Oxygen* (H. H. Wasserman and R. W. Murrany, eds.), Academic Press, New York (1979).
31. H. Nakazumi, E. Hamada, T. Ishiguro, H. Shiozaki, and T. Kitao, *J. Soc. Dyers and Colourists 105*, 26 (1989).
32. S. Yasui, M. Matsuoka, and T. Kitao, *Dyes Pigments 10, 13* (1989).
33. K. Namba, Preprint of the Annual Meeting of the Chemical Society of Japan, 4D411, Sendai (1988).

7

Photochromic Dyes

JUN'ETSU SETO

I. INTRODUCTION

Photochromism is the phenomenon whereby the absorption spectrum of a molecule or crystal changes reversibly when the sample is irradiated by light of certain wavelengths. A colorless compound **A**, for instance, changes its molecular structure to a quasi-stable colored structure **B** when irradiated by ultraviolet (UV) light; **B** can be returned to the colorless structure **A** by exposure to visible light or heating.

A large number of organic and inorganic materials which exhibit photochromism have been known for many years, and such materials have found application in many areas including as dosimeter materials, light control filters, recording films in printing processes, and decorative paints. Recently, these types of materials have attracted much attention in the fields of optical recording media and "molecular electronic devices." This is largely because these materials have potential as ultra-high-density information storage media, since the photochromic reaction occurs at the molecular level, especially in organic compounds.

Although their application as such erasable recording media has been expected, the photosensitive wavelengths of the photochromic materials studied so far have been in the visible region (400–700 nm), and these materials do not have significant sensitivity in the infrared region near 800 nm, the wavelength region of conventional laser diodes. The develop-

JUN'ETSU SETO • Research Center, SONY Corporation, Hodogaya-ku, Yokohama, Kanagawa 240, Japan.

ment of new photochromic compounds that are sensitive to wavelengths around 800 nm is eagerly awaited.

We thus conducted research to improve the properties of spiropyran compounds and recently found a new class of spiropyrans with the required sensitivity in the infrared region.

In this chapter, the synthesis, photochemical properties, and application to optical recording media of the new spiropyrans are described following a review of typical photochromic compounds.

II. TYPICAL PHOTOCHROMIC COMPOUNDS

In recent years, the synthesis of new photochromic compounds and the elucidation of their structures and photochemical properties have been the subject of active study. Some typical photochromic compounds are listed in Table I.

The spiropyran compound in the table is a well-known material which we have made suitable for use as an optical recording material by synthesizing appropriate derivatives.[1] Heller and co-workers synthesized stable fulgide compounds and were the first to apply them to recording media.[2,3] Tashiro and Yamamoto investigated the effect of the polymer matrix on the photochromic properties of a newly synthesized dihydropyrene compound.[4] The application of thioindigo compounds to photoimage sensors was studied by Takahashi et al.[5] The colorless cis form was converted to the colored trans form by the image focuses onto the detector, and the luminescence produced by the trans form upon irradiation with a He–Ne laser was read. Dürr et al. synthesized a bipyridine bichromophoric compound for use in display devices and light control filters.[6] Irie recently synthesized dimethylthiophene derivatives with remarkably high thermal stability.[7]

Many other photochromic compounds such as aziridine, oxazine, azobenzene, salicylideneaniline, xanthene, and polynuclear aromatic derivatives[8,9] have been studied.

This review of the typical photochromic compounds has focused mainly on the active areas of study. For application of these materials, the appropriate choice of a compound with the desired coloration wavelength, determined by the application, is important.

III. SYNTHESIS OF IR-SENSITIVE SPIROPYRAN

At present, the shortest wavelength of laser diodes in practical use is 780 nm, and the shortest wavelength that can be expected in the near future

Table I. Typical Photochromic Compounds[a]

Type of compound	Reactions involved in photochromic behavior
Spiropyran	$\xrightleftharpoons[h\nu_2, \Delta]{h\nu_1}$
Fulgide	$\xrightleftharpoons[h\nu_2]{h\nu_1}$
Dihydropyrene	$\xrightleftharpoons[h\nu_2, \Delta]{h\nu_1}$
Thioindigo	$\xrightleftharpoons[h\nu_2]{h\nu_1}$
Bipyridine	$\xrightleftharpoons[\Delta]{h\nu_1}$
Aziridine	$\xrightleftharpoons[\Delta]{h\nu}$
Polynuclear aromatics	$\xrightleftharpoons[h\nu(313nm)]{h\nu(405nm)}$
Azobenzene	$\xrightleftharpoons[h\nu_2]{h\nu_1}$
Salicylideneaniline	$\xrightleftharpoons[h\nu_2]{h\nu_1}$
Xanthene	$\xrightleftharpoons[h\nu_2]{h\nu_1}$
Oxazine	$\xrightleftharpoons[h\nu_2]{h\nu_1}$

[a] Ref. 1.

is 680 nm. Therefore, in order to use photochromic materials as erasable recording media, new photochromic compounds which absorb at longer wavelengths comparable to that of the laser diode must be synthesized.

A. Synthetic Method

An attempt to obtain a long-wavelength absorption for the merocyanine form, the colored form of spiropyran, by synthesizing a compound containing the benzothiopyran ring was first made by Becker and Kolc[10] (Scheme 1). They showed that the colored form of 1',3',3'-trimethylspiro[2H-1-benzothiopyran-2,2'-indoline] had a strong absorption band in the 600–850-nm wavelength region in 3-methylpentane at 77 K, but the merocyanine form was not produced at room temperature either in solution or in a polymer film.

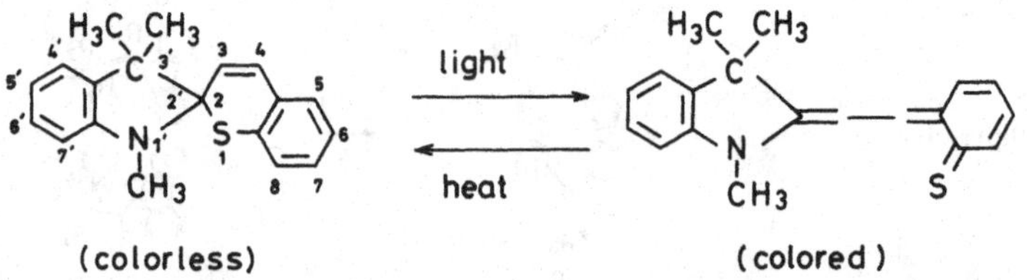

Scheme 1

In order to obtain photochromic compounds whose colored forms have a strong absorption band in the 700–850-nm wavelength region and are stable even at room temperature, our research group synthesized indolinospirobenzothiopyrans with a nitro group at the 6-position.[11] Indolinospirobenzothiopyrans 1 with a nitro group at the 6-position were synthesized by the reaction of 1,3,3-trimethyl-2-methyleneindolines 2 and 5-nitrosalicylaldehydes 3 in ethanol (Scheme 2).

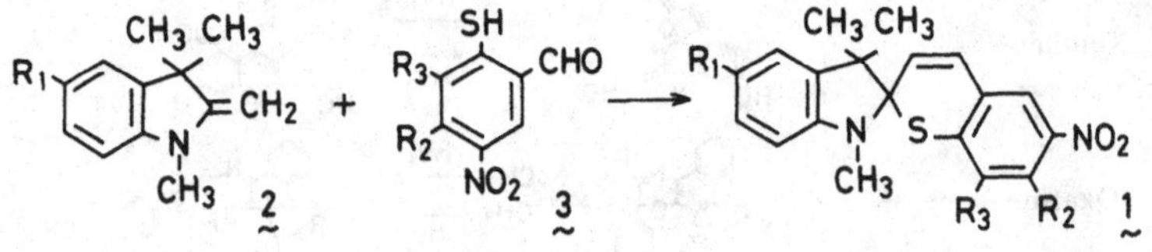

Scheme 2

The compounds **2** are prepared according to the reaction in Scheme 3, by methylation of 2,3,3-trimethylindolenines **4** at the N-position with methyl iodide to give indolenium salts **5**, which are subsequently treated with base.

The compounds **4** were obtained through Fischer's indole synthesis, in which substituted phenylhydrazines or their hydrochloride salts and 3-methyl-2-butanone are heated under acidic conditions (Scheme 3).[12]

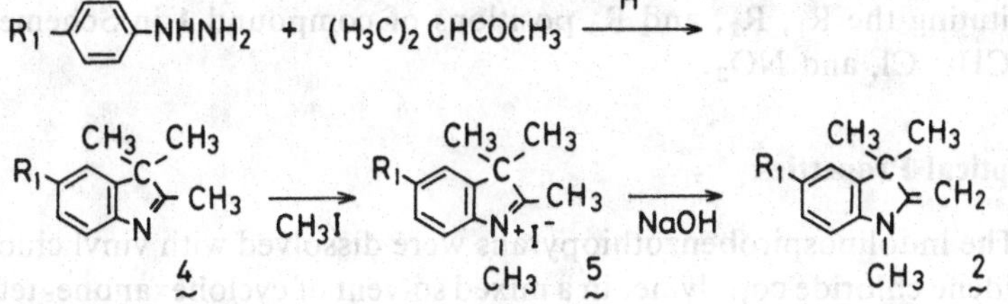

$$R_1 - \langle phenyl \rangle - NHNH_2 \ + \ (H_3C)_2 CHCOCH_3 \ \xrightarrow{H^+}$$

$$\underset{4}{R_1 - \text{indolenine}} \ \xrightarrow{CH_3I} \ \underset{5}{R_1 - \text{indolenium}} \ \xrightarrow{NaOH} \ \underset{2}{R_1 - \text{methylene indoline}}$$

Scheme 3

The 5-nitrothiosalicylaldehydes **3** were prepared by the following two-methods: (A) conversion of a halogen atom of o-chlorobenzaldehyde into an SH group,[13] and (B) conversion of an OH group of salicylaldehyde into an SH group (Scheme 4).[14]

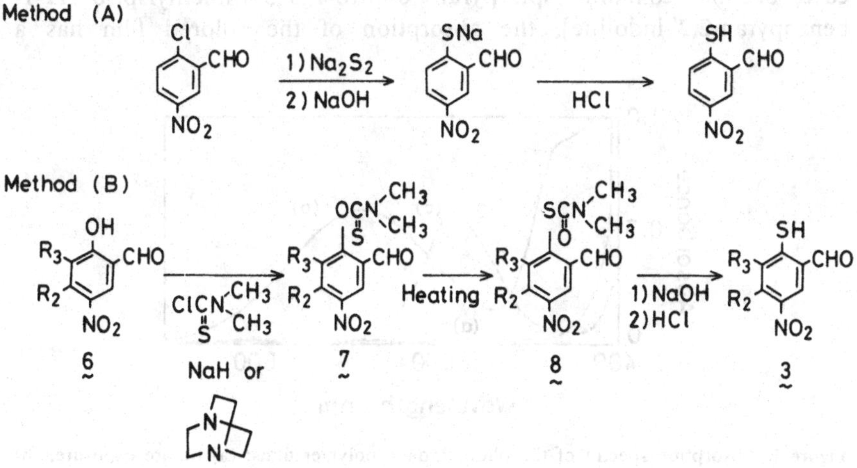

Method (A)

$$\underset{NO_2}{Cl-C_6H_3-CHO} \ \xrightarrow[2)\,NaOH]{1)\,Na_2S_2} \ \underset{NO_2}{SNa-C_6H_3-CHO} \ \xrightarrow{HCl} \ \underset{NO_2}{SH-C_6H_3-CHO}$$

Method (B)

$$\underset{6}{\underset{NO_2}{R_3,R_2-OH,CHO}} \ \xrightarrow[\substack{NaH\ or}]{ClCN(CH_3)_2\,C=S} \ \underset{7}{\underset{NO_2}{R_3,R_2-OCN(CH_3)_2\,(C=S),CHO}} \ \xrightarrow{Heating} \ \underset{8}{\underset{NO_2}{R_3,R_2-SCN(CH_3)_2\,(C=O),CHO}} \ \xrightarrow[2)\,HCl]{1)\,NaOH} \ \underset{3}{\underset{NO_2}{R_3,R_2-SH,CHO}}$$

Scheme 4

Method A could be applied only to the preparation of 5-nitrothiosalicyl-aldehyde, because the starting o-chlorobenzaldehydes with various sub-stituents were not readily available. On the other hand, method B is excellent for the preparation of 5-nitrothiosalicylaldehyde derivatives, because the starting salicylaldehydes are easily prepared and the yield in each reaction is very high. In Method B, dimethylthiocarbamates **7** were first prepared, followed by heating for the rearrangement reaction of O and S to give compounds **8** and subsequent alkaline hydrolysis to give thiosalicylaldehy-des **3**.

Many indolinospirobenzothiopyran derivatives were synthesized by substituting the R_1, R_2, and R_3 positions of compound **1** in Scheme 2 by H, OCH_3, Cl, and NO_2.

B. Optical Properties

The indolinospirobenzothiopyrans were dissolved with vinyl chloride-vinylidene chloride copolymer in a mixed solvent of cyclohexanone-tetrahy-drofuran. The resulting solution was spin-coated onto a quartz glass plate and dried to obtain a photosensitive film of around 1-μm thickness.

On exposure to UV light (360 nm), the light yellow film changed to dark green in color. The absorption spectrum of the spirobenzothiopyran **1** ($R_1 = R_2 = R_3 = H$) is shown in Figure 1. The nonirradiated film has no absorption in the wavelength region of 500–900 nm (Figure 1a), but the irradiated film has an absorption maximum (λ_{max}) at 680 nm and the absorption extends to about 900 nm (Figure 1b). On the other hand, in the case of the common spirypyran 6-nitro-1',3',3'-trimethylspiro[2H-1-benzopyran-2,2'-indoline], the absorption of the colored film has a

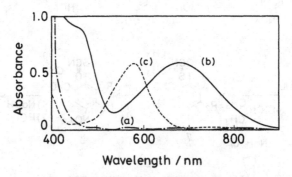

Figure 1. Absorption spectra of the photochromic polymer films: (a) before exposure, (b) after exposure to UV light of a film containing spirobenzothiopyran **1**, and (c) after exposure to UV light of a film containing a spiropyran of the conventional type.[11]

maximum at 580 nm and the absorption on the long-wavelength side extends only to 700 nm (Figure 1c). Thus, it was found that the substitution of sulfur for oxygen at the 1-position of spiropyran caused dramatic red shifts of more than 100 nm.

The effect of substituents on the absorption maxima in the colored polymer films was also investigated. It was found that using Brown's σ_p^+ relationship for substituents at the 5'-positions of the indolinic ring (σ_i) and Brown's σ_m^+ relationship and Taft's σ^* values substituents at the 7- and 8-positions, respectively, of the benzothiopyran moiety (σ_b),[15] a good correlation between λ_{max} and the substituent constants was obtained. That is, as σ_i increases or σ_b decreases, the absorption maximum shifts to longer wavelengths.

The stability of the colored forms was improved by varying the R_1-R_3 substituents of the spirobenzothiopyran 1, so that a half-life for the colored state at 40°C of more than one month was obtained.

The absorption maximum of the colored form of the new spirobenzothiopyran in ethanol is at 620 nm ($\varepsilon = 2.5 \times 10^3 \, dm^3 \, mol^{-1} \, cm^{-1}$). The value of the molar absorptivity is larger than that expected for $\pi-\pi^*$ transitions. The absorption shows a blue shift with increasing solvent polarity similar to that of the ordinary spirobenzopyrans. These findings indicate that the assignment of the absorption band of the new spirobenzothiopyran is the same as that for the ordinary spirobenzopyrans.

Quantum-chemical studies have been made on the colored forms of ordinary spirobenzopyrans. The CNDO/S method has been applied to the four *trans* isomers of the colored form. The calculated results show that the zwitterionic structure is characteristic of the colored form, and the absorption spectrum involves the $\pi-\pi^*$ transition associated with charge transfer from the indoline ring.[16] This assignment should also hold for the absorption band of the new compound.

IV. PHOTOCHROMIC PROPERTIES OF SPIROBENZOTHIOPYRAN[17]

A. Coloration and Decoloration Process

The photochemistry of the coloration and decoloration process was investigated in aerated and degassed solutions of ethanol (EtOH) and N,N-dimethylformamide (DMF). Excitation was from a pulsed dye laser (FWHM > 8 ns; output, 10 mJ) optically pumped by a XeCl excimer laser (see block diagram in Figure 2). The absorption change was detected by a K-367 laser kinetic spectrometer (Applied Photophysics). The photomulti-

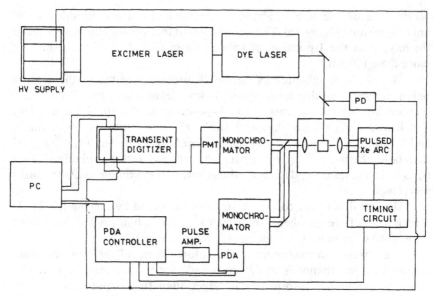

Figure 2. Block diagram of laser-flash apparatus.

plier signal was recorded on a Tektronix 7912AD programmable digitizer and analyzed by a personal computer.

The colorless form of the spirobenzothiopyran compound was excited at 360 nm. Figure 3 shows the transient absorption spectra in aerated and degassed EtOH solutions, measured immediately after the laser pulse. There is no significant difference between the two spectra. The shape of the spectrum was that of the colored form, and no reaction intermediates were detected. Similar results were obtained in aerated and degassed DMF solutions.

Kinetic measurements of the coloration process were made at the wavelength of the absorption maximum of the colored form (620 nm in EtOH, 660 nm in DMF). The absorbance of the colored form was proportional to the laser energy, indicating a single-photon process. Figure 4 shows the result of the kinetic measurements of the coloration process in EtOH. The time that the absorption change lags behind the time-integrated laser intensity is very small. We assumed a linear response of the reaction to the laser pulse and obtained a good fit of the experimental results with the calculated absorption changes using various rate constants. The coloration proceeds by first-order kinetics, and the coloration rates were determined to be $7 \times 10^7 \, s^{-1}$ in EtOH and $8 \times 10^7 \, s^{-1}$ in DMF. No significant difference was found in the observed rate constants for degassed and aerated solutions or for aprotic and protic solvents.

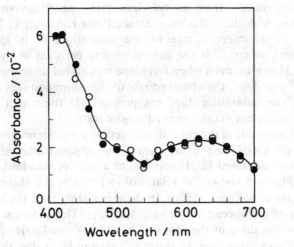

Figure 3. Transient absorption spectrum observed 30 ns after a laser pulse in aerated (○) and degassed (●) EtOH solutions containing spirobenzothiopyran **1**.[17]

The reaction state of the coloration process was determined through triplet quenching experiments on the photoexcited colorless form in degassed EtOH solution using ferrocene [triplet energy, 159 kJ mol^{-1} (ref. 18)] as a quencher. It is reasonable to assume that the triplet energy level of the colorless form is much higher than that of ferrocene, so the triplet-

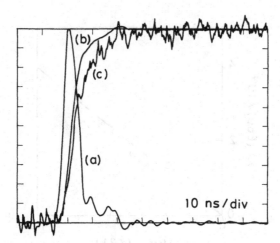

Figure 4. The photocoloration process for spirobenzothiopyran **1** in ethanol solution. The laser profile (a), its time-integrated intensity (b), and the absorption change of the colored form (c) are shown. Each curve is normalized to unity.[17]

triplet energy transfer from the colorless spirobenzothiopyran to ferrocene may proceed with the diffusion-controlled rate constant, k_d [$9.2 \times 10^7 \text{ s}^{-1}$ (ref. 19)]. The observed rate of the coloration, k, is given by $k = k_0 + k_d[\text{Ferr}]$, where k_0 is the coloration rate constant in the absence of ferrocene. However, even when ferrocene was added up to a concentration of $1 \times 10^{-1} \text{ mol dm}^{-3}$, the observed rate of the coloration was not changed. Therefore, the coloration does not proceed via the triplet state of the uncolored form, but via the excited singlet state.

Quantum yields of the coloration reaction were determined in degassed solutions of EtOH and DMF using the triplet–triplet absorption of anthracene in degassed EtOH solution as a relative standard as described in ref. 5. Figure 5 shows the relationship between the absorption of the colored form at 660 nm, $A(660)$, in degassed DMF and the triplet–triplet absorption of anthracene at 421 nm, $A_{\text{trip}}(421)$. The quantum yield is estimated from the slope of the plot and is 4.8×10^{-2} in EtOH and 1.8×10^{-2} in DMF. These values are an order of magnitude smaller than those for the ordinary spirobenzopyrans,[20] which indicates that the deactivation process due to the sulfur atom is efficient for the new spiro compound.

For the decoloration reaction, the colored form was excited in the wavelength region on the longer-wavelength side of the absorption spectrum by a laser flash. In aerated solutions, decoloration was observed, while in degassed solutions, no decoloration of the colored form was observed under these experimental conditions. Therefore, the photodecoloration observed in aerated solution was due to photooxidation, and the conversion of the

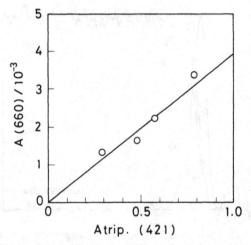

Figure 5. Plot of the absorbance of the colored form of spirobenzothiopyran **1** in degassed DNF, $A(660)$, versus the absorbance measured for the triplet–triplet absorption of anthracene, $A_{\text{trp}}(421)$.[17]

colored form to the colorless form occurs mainly by way of a thermal
reaction in the ground state.

B. Solvent Effects

The thermal stability of the colored form is affected by the solvent
polarity. A number of studies have been conducted on the ordinary spirobenzopyrans, and the colored forms are best regarded as zwitterionic species.[21]
The logarithm of the fading rate constant decreases linearly with increasing
Kosower's Z or Brownstein's S values. However, the plot of the logarithm
of the fading rate constant against $(D - 1)/(D + 2)$, where D is the dielectric
constant of the solvent, is nonlinear and is different for protic and aprotic
solvents. These results can be interpreted in terms of hydrogen bonding
between the colored form and protic solvents.

For the new spirobenzothiopyran, the thermal fading rate, k_T, at 25°C
and the wavenumber of the absorption maximum of the colored form was
measured in methanol (MeOH), EtOH, isobutyl alcohol (i-BuOH), tetrahydrofuran (THF), DMF, and dimethyl sulfoxide (DMSO). Remarkably
different solvent effects for the thermal fading rate were observed for this
compound. Figure 6 shows the plot of the logarithm of k_T against Dimroth's
E_T values[22] (E_T is linearly related to the S and Z values) and $(D - 1)/(D + $

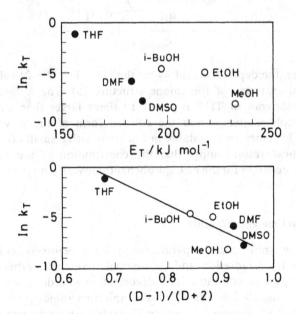

Figure 6. Plots of ln k_T versus E_T (upper half) and $(D - 1)/(D + 2)$ (lower half) for the colored
form of spirobenzothiopyran **1** in aprotic solvents (●) and protic solvents (○).[17]

2). With increasing E_T and $(D - 1)/(D + 2)$, k_T decreases. A linear relationship was obtained against $(D - 1)/(D + 2)$, rather than against E_T, which is contrary to the relationship found for the ordinary spirobenzopyrans. For the new spirobenzothiopyran, the hydrogen bonding effect on the thermal stability of the colored form in protic solvents, observed for the ordinary spiropyrans, was not marked.

For the structure of the colored form of the new spirobenzothiopyran, the two resonance forms shown in Scheme 5, the zwitterionic structure (b) and the thione structure (a), are considered. Becker and Kolc[10] assigned the colored form of a spirobenzothiopyran a similar structure to (a), by comparing the spectral characteristics at 77 K with those of the colored forms of the chromenes, $2H$-1-benzothiopyrans, and spiro[benzopyran-2,2'-indolines].

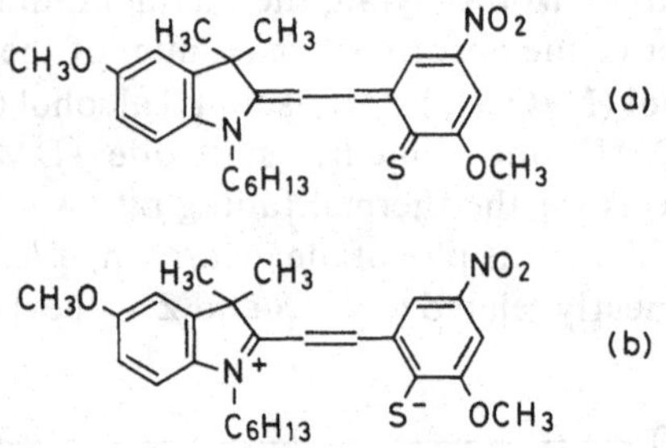

Scheme 5

However, the dependence of k_T on the $(D - 1)/(D + 2)$ value cannot be explained in terms of the thione structure (a). The k_T for the new spirobenzothiopyran in THF is about 10 times larger than that for the ordinary spirobenzopyran, and the colored form of the new spirobenzothiopyran becomes very unstable in a solvent with a small $(D - 1)/(D + 2)$ value. These results imply that the contribution of the zwitterionic structure is even larger for the new spirobenzothiopyran than for the ordinary spirobenzopyran.

C. Cycle Lifetime and Stability

The new spirobenzothiopyran could be repetitively colored and bleached by UV irradiation and by thermal treatment, respectively. We performed repetitive coloration and decoloration in a degassed DMF solution containing 10^{-4} mol dm^{-3} of the spirobenzothiopyran. With two minutes of UV irradiation, more than 30 cycles were achieved before the

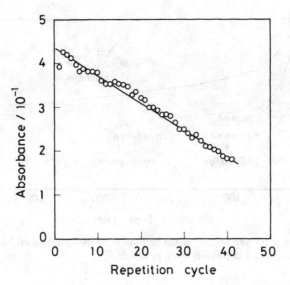

Figure 7. Decay of the absorption of the colored form of spirobenzothiopyran 1 on repetitive UV coloration and thermal fading cycles.[17]

absorption maximum of the colored form decreased to 50% of the initial value (Figure 7). The value of the cycle lifetime is on the order of 10 even in a dilute solution, which is promising for practical application.

The photochromic materials are usually used in the form of a solid film made by solvent casting a homogeneous solution of the photochromic compound dissolved in a polymeric material. The thermal stability of the coloration in this photochromic film is affected greatly by the properties of the polymeric materials. As expected from the previous "solvent effect" results, the thermal stability of coloration is improved by the use of polar polymers, such as poly(vinyl chloride), vinyl chloride/vinylidene chloride copolymer, and acrylic polymers.

Moreover, the addition of phenol compounds markedly enhances the stability.[1] Figure 8 shows that the bisphenol A type is the most effective and gives 10 times greater stability than other phenol compounds, such as simple phenols and hindered phenols which are used as antioxidants for polymers. It seems that the spatial arrangements of the hydroxyl groups play an important role.

D. Related Materials (Spiro-oxazine)

Recently, another interesting spiro compound, indolinonaphthoxazine, has been reported. This compound is similar to the conventional spiropyrans

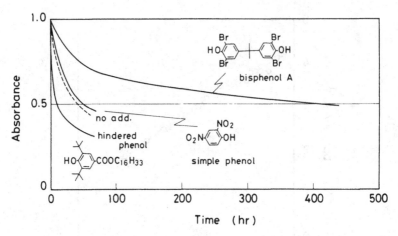

Figure 8. Effects of the addition of phenol derivatives on the bleaching of the colored form of spirobenzothiopyran 1 in polymer films, at 30°C.[1]

in its structure and photochemical reactions. The photochromic cyclability is reported to be better than for spiropyran, but a compound with an absorption band which extends into the infrared region has yet to be reported.

The spiroindolinonaphthoxazine is readily prepared by refluxing 1-nitroso-2-naphthol with a 3,3-dimethyl-2-methyleneindoline derivative in alcohol (Scheme 6).

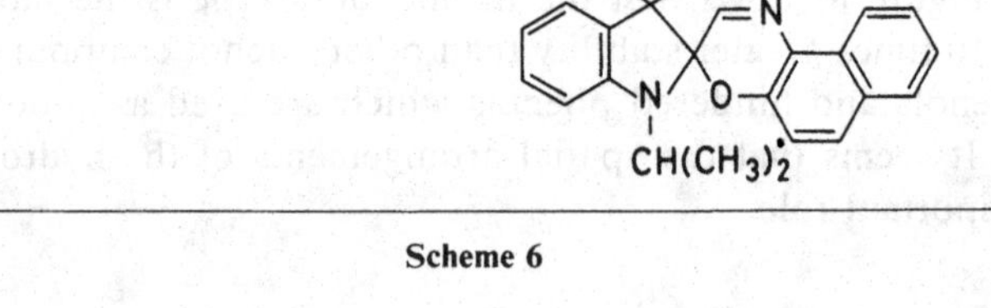

Scheme 6

The photochemical properties of the N-methyl derivatives[23] have been studied. The colorless form in ethanol has three absorption bands (203, 235, 320 nm) and is colored upon irradiation with UV light to give the

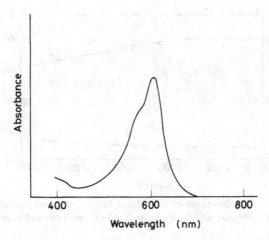

Figure 9. Absorption spectrum of spiroindolinonaphthoxazine in EtOH solution.[23]

absorption spectrum shown in Figure 9. The spectrum has an absorption band in the visible region, with a maximum at 612 nm ($\varepsilon = 8.1 \times 10^4 \, dm^3 \, mol^{-1} \, cm^{-1}$) and a shoulder at 578 nm ($\varepsilon = 4.9 \times 10^4 \, dm^3 \, mol^{-1} \, cm^{-1}$). This absorption is assigned to the π-π^* transition associated with charge transfer from the indoline to the naphthalene ring.

The thermal decay of the colored form is fairly rapid, with a rate constant of $2.1 \times 10^{-1} \, s^{-1}$ at 20°C. The absorption maximum shows a slight red shift with an increase in the solvent polarity,[24] which is opposite to the shift in the spiroindolinobenzopyrans. The structure of the colored form is assigned to a keto type, which is consistent with INDO/S–CI calculations.

Spiroindolinonaphthoxazine shows good durability with respect to repetitive coloration and decoloration.[25] In tests where an isopropyl alcohol solution with an initial absorbance of 0.26 was repetitively irradiated with first an XeCl excimer laser (308 nm) and then visible light ($\lambda > 450 \, nm$), the absorption decreased by only 20% after 5×10^3 cycles, which corresponds to a 5% change in transmittance (Figure 10). For a sample with an initial absorbance of 1, the transmittance increases only by 4% after 1×10^3 cycles.

The cycle lifetime was affected by the solvent employed; for example, the number of cycles in which the absorption decreases to 50% of the initial value was 4×10^2 in acetonitrile and 2.3×10^3 in benzene. However, the mechanism of the solvent effect has not as yet been elucidated.

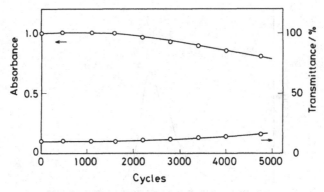

Figure 10. Decay of the absorption of the colored form of spiroindolinonaphthoxazine on repetitive cycles of coloration and fading by an XeCl excimer laser and visible light, respectively.[25]

V. APPLICATION TO OPTICAL RECORDING MEDIA[26]

A new class of photochromic spiropyrans with benzothiopyran units in the molecular framework was described in the previous section. The absorption spectrum of a polymer film containing such a material is shown in Figure 1. The nonirradiated film has no absorption in the wavelength region of 500–900 nm, but when irradiated with UV light, the film becomes dark green in color with high absorption in the 600–850-nm region.

These interesting properties of spirobenzothiopyran are well suited to the requirements of erasable optical recording media for systems using conventional laser diodes.

The structure of the recording media is as follows. The medium consists of three thin layers: a recording layer, a highly refracting layer, and a reflective layer, as shown in Figure 11. The recording layer is formed by spin coating (thickness, 0.7 μm) and includes a photochromic material, a

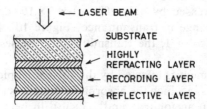

Figure 11. Structure of the optical recording medium.[1]

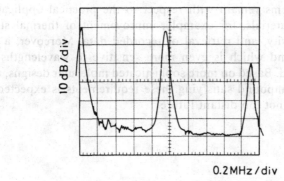

0.2 MHz / div

Figure 12. Carrier-to-noise ratio (CNR) for the disk medium measured by a spectrum analyzer.[1]

polymer binder, and various additives. The highly refracting layer is prepared by the vacuum deposition of zinc sulfide (ZnS; refractive index, 2.3) onto a glass substrate. This layer (thickness, 850 Å) serves to increase the sensitivity of the recording medium. That is, the optimized thickness of these layers results in a very low reflectivity from the substrate so that most of the incident beam is absorbed in the recording medium. The reflective layer (thickness, 1500 Å) consists of vacuum-deposited silver, which has high reflectivity at the laser-diode wavelength.

The recording properties of the medium incorporating these materials were examined. Before recording, the medium was exposed to UV light and changed to the colored form. Optical recording on the medium was then carried out with a laser diode (wavelength, 780 nm; power, 7.5 mW). The laser beam was focused to a 2-μm-diameter spot. The sensitivity of recording was 0.15 J cm^{-2} for a 20% change of reflectivity. The carrier-to-noise ratio (CNR) was more than 54 dB in a 30-kHz bandwidth at 1-MHz carrier frequency, as shown in Figure 12. These recording characteristics show that the material acts as a high-quality erasable optical medium for use with laser diode optical systems.

VI. SUMMARY AND CONCLUSION

The recent research activity on photochromic materials in general has been reviewed. With regard to the infrared absorbing photochromic compounds, an in-depth account of the work that we have carried out on newly developed benzothiopyran compounds has been given, covering the synthetic method, the photochemical properties of these compounds, and their application as optical recording media.

Many problems remain with respect to the practical application of photochromic materials, for example, improvement of thermal stability, shelf life, sensitivity, and retrieval of recorded data. Moreover, a photochromic compound which is even more sensitive to wavelengths above 680 nm is required. Based on more sophisticated molecular designs, a novel photochromic compound satisfying these requirements is expected to be developed in the not too distant future.

REFERENCES

1. J. Seto, *Solid State Phys. 21*, 235 (1986).
2. H. G. Heller and R. J. Langan, *J. C. S. Perkin Trans. 2*, 341 (1981).
3. H. G. Heller, *IEE Proc. 130*(1), 209 (1983).
4. M. Tashiro and T. Yamamoto, *J. Org. Chem. 47*, 2783 (1983).
5. T. Takahashi, Y. Taniguchi, K. Umetani, M. Hashimoto, and T. Kano, *Jpn. J. Appl. Phys. 24*, 173 (1985).
6. H. Dürr, P. Spang, and G. Huack, Proceedings of the Xth IUPAC Symposium on Photochemistry, p. 363 (1984).
7. M. Irie, *Kagaku to Kogyo 41*, 163 (1988) (in Japanese).
8. A. M. Trozzolo, M. A. Zottola, and L. Riu, Proceedings of the Xth IUPAC Symposium on Photochemistry, p. 203 (1984).
9. R. Schmidt, W. Drews, and H. D. Brauer, *J. Phys. Chem. 86*, 4909 (1982).
10. R. S. Becker and J. Kolc, *J. Phys. Chem. 72*, 997 (1968).
11. S. Arakawa, H. Kondo, and J. Seto, *Chem. Lett.*, 1805 (1985).
12. B. Robinson, *Chem. Rev. 63*, 373 (1963).
13. C. C. Price and G. W. Stacy, *J. Am. Chem. Soc. 68*, 498 (1986).
14. M. S. Newman and H. A. Karnes, *J. Org. Chem. 31*, 3980 (1966).
15. P. Berman, R. E. Fox, and F. D. Thomson, *J. Am. Chem. Soc. 81*, 5605 (1959).
16. V. N. Lisyutenok and V. A. Barachevskii, *Teor. Eksp. Khim. 17*, 485 (1981).
17. J. Seto, S. Arakawa, H. Kondo, S. Tamura, and N. Asai, Abstracts of the XIIth International Conference on Photochemistry, p. 105, Tokyo, Japan (1985); *Bull. Chem. Soc. Jpn. 62*, 358 (1989).
18. A. Farmilo and F. Wilkinson, *Chem. Phys. Lett. 34*, 575 (1975).
19. L. Murov, *Handbook of Photochemistry*, p. 55, Marcel Dekker, New York (1973).
20. S. Kholmanskii, A. V. Zubkov, and K. M. Dyunaev, *Russ. Chem. Rev. 50* (1981).
21. C. Bertelson, *Photochromism*, p. 171, Wiley-Interscience, New York (1971).
22. C. Reichardt, *Angew. Chem. Int. Ed. Engl. 4*, 29 (1965).
23. Nori, Y. C. Chu, *Can. J. Chem. 61*, 300 (1983).
24. S. Nakamura, Proceedings of 24th Meeting, The Japanese Association for Organic Electronic Materials, p. 7 (1988).
25. M. Irie and K. Hayashi, *Polymer Preprints, Japan 34*, 459 (1985).
26. S. Arakawa, H. Kondo, S. Tamura, N. Asai, and J. Seto, Technical Digest, Conference on Lasers and Electro-Optics, p. 358, San Francisco (1986).

8

Other Chromophores

MASARU MATSUOKA

I. INFRARED ABSORBING AZO DYES

There are many dye chromophores which absorb in the near-infrared region. These are classified as cyanines, quinones, phthalocyanines, and 2:1 metal complex dyes, as described in the previous chapters. Azo dyes are the most popular chromophore but generally absorb in the visible region. The well-known deep-colored azo disperse dyes have the general formula **1** and absorb below 650 nm. Recently, Bello and Griffiths[1] designed new

$$A = NO_2, CN$$
$$D = NHAc, OR$$

(1)

types of infrared absorbing azo dyes which absorb above 700 nm. They have a chromophoric system of strong intramolecular charge-transfer (CT) character. A large bathochromic shift can be attained by the introduction of a heteroaromatic ring (thiazole) in place of a benzenoid ring, and additional substitution of the strong acceptor produced a large bathochromic shift in λ_{max} and an increase of molar absorptivity caused by the enlargement

MASARU MATSUOKA • Department of Applied Chemistry, College of Engineering, University of Osaka Prefecture, Sakai, Osaka 591, Japan.

Table I. Substituent Effects on λ_{max} and ε of 2^a

(2)

	X	R^1	R^2	R^3	R^4	$\lambda_{max}^{\ b}$ (nm)	$\Delta\lambda$	$\log \varepsilon$
2a	CHO	Et	Et	H	H	574	—	4.71
2b	CH=C(CN)$_2$	Et	Et	H	H	645	71	4.79
2c	C(CN)=C(CN)$_2$	Et	Et	H	H	725	151	4.88
2d	(Y = C=O)	Et	Et	H	H	700	—	4.83
2e	Y=SO$_2$	H	R^c	OMe	NHAc	778	78	4.92

^{a} Ref. 1.
^{b} Measured in dichloromethane.
^{c} R^2 = CH(Me)Bu(n).

of the π-conjugated system. The dependence of the λ_{max} and ε values of the azo dyes on the strength of the donor–acceptor system is shown in Table I.

Substitution of a cyano group produced a large bathochromic shift in λ_{max} from **2a** to **2c**, and the replacement of a carbonyl group in **2d** by a sulfone in **2e** also produced a large bathochromic shift. The bathochromic shift in λ_{max} was accompanied by an increase in the ε values; thus, a large bathochromic shift and an increase in the ε value can be obtained with an increase in the intramolecular CT character.

It is generally known that the λ_{max} of azo dyes shifts to longer wavelengths with an increase in the number of azo units, up to a certain number of azo units.[2] On the other hand, in some cases of polyazo dyes with intramolecular CT chromophores, disazo dyes absorb at much longer wavelengths than the corresponding trisazo dyes containing the same donor and acceptor groups.[3] Disazo dye **3** absorbs at 565 nm but trisazo dye **4** absorbs at 545 nm.

Disazo dye **5** absorbs at 710 nm; structural modification by means of substituent effects and/or annelation to the phenyl conjugating unit effectively produced the bathochromic shift in λ_{max}. It is now possible to obtain new infrared absorbing monoazo and polyazo dyes which have special characteristics for various applications.

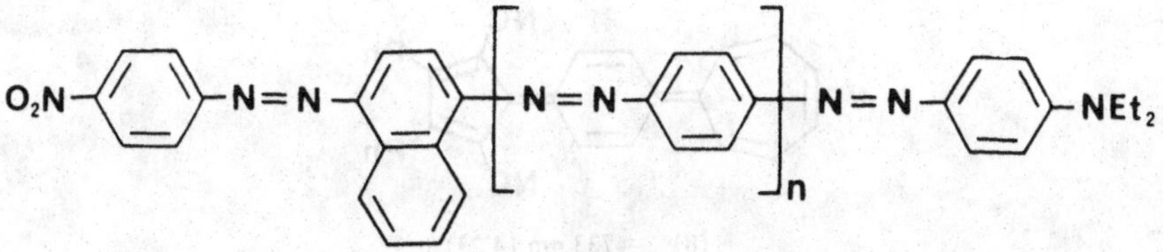

(3) $n = 0$

(4) $n = 1$

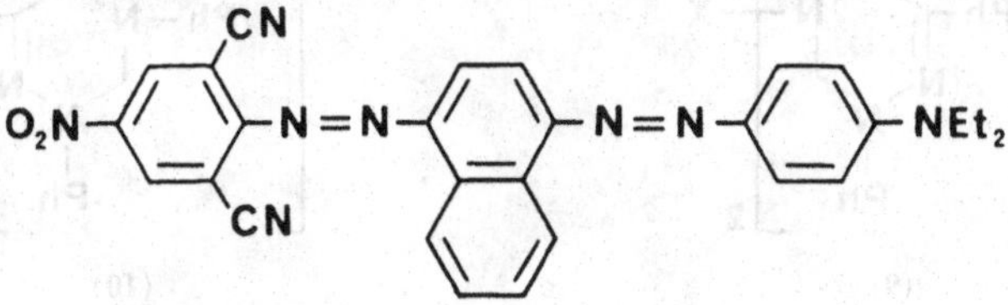

(5) 710 nm, log ε = 4.64

II. MISCELLANEOUS CHROMOPHORES

Intramolecular CT chromophores can be used to produce many types of infrared absorbing dyes. Strong donor–acceptor systems such as dyes 6[4] and 7 are typical examples, and they absorb at 735 and 755 nm, respectively. A bathochromic shift in λ_{max} can be attained by introducing much stronger donors and/or acceptors in these chromophoric systems.

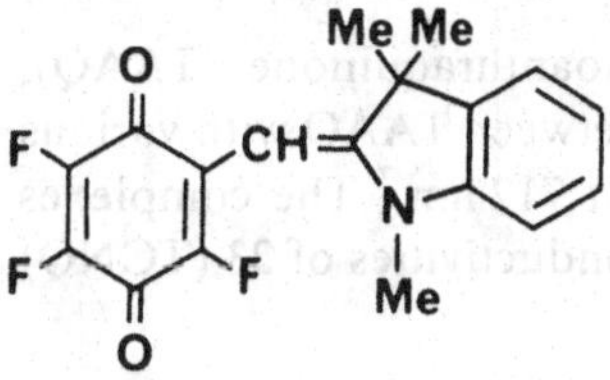

(6)   735 nm (4.16)

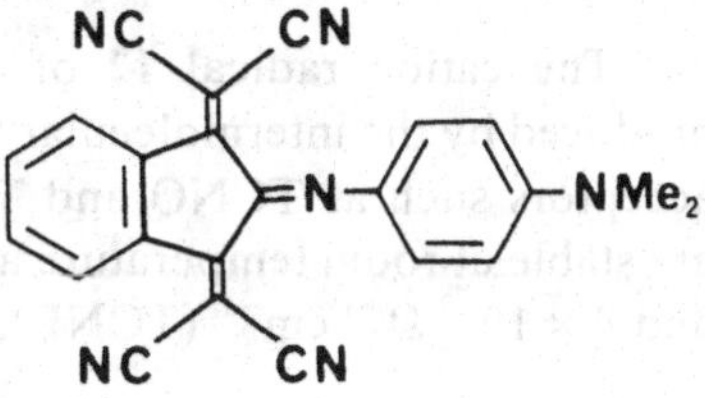

(7)   755 nm

On the other hand, nonbenzenoid aromatic chromophores such as 8 absorb at longer wavelengths.[5] The tropone moiety acts as a strong π donor and is stabilized by a 6π-electron system. The chromophore produces the highly delocalized π-conjugated system to give infrared absorption.

The tetrazine radicals 9 and 10, and their cation salts 11, absorb in the infrared region,[6] but their stability as coloring agents are unknown.

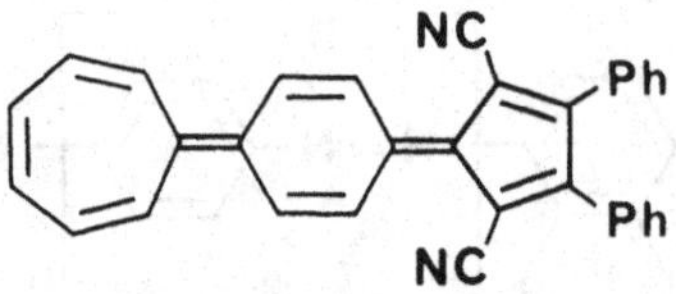

(8)   733 nm (4.23)

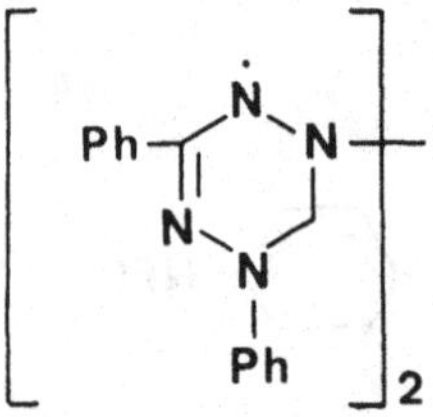

(9)

X = p-phenylene   773 nm (4.42)
X = p-terphenyl   758 nm (4.15)

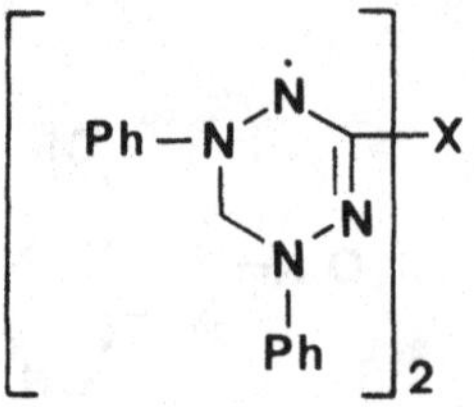

(10)

X = p-biphenyl   762 nm (4.13)

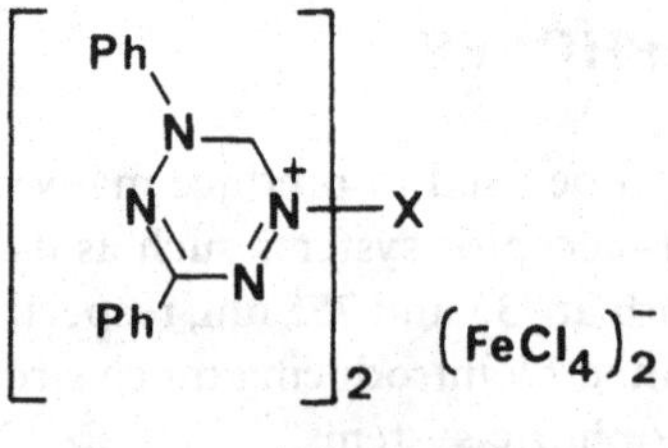

(11)

X = p-phenylene   885 nm (4.34)

The cation radical **12** of 1,4,5,8-tetraaminoanthraquinone (TAAQ), produced by the intermolecular CT interaction between TAAQ with various acceptors such as TCNQ and TCNE, absorbs at 712 nm.[7] The complexes are stable at room temperature and have good conductivities of 23 (TCNQ) and $4 \times 10^{-3} \, \Omega^{-1} \, cm^{-1}$ (TCNE), respectively.

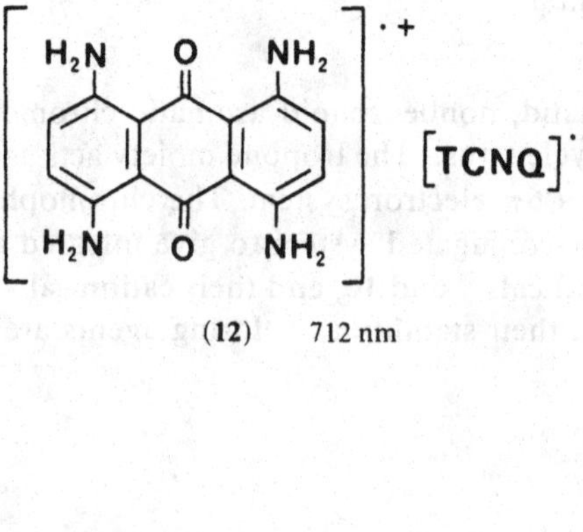

(12)   712 nm

III. INTERMOLECULAR CHARGE-TRANSFER COMPLEX DYES

There are many dye chromophores which show intramolecular CT absorption spectra, as described in the previous chapters, but few dyes containing intermolecular CT chromophores are used practically as coloring agents.

The relation between color and constitution of intermolecular CT dyes was recently evaluated quantitatively by means of the PPP MO method.[8] Matsuoka et al.[9] synthesized infrared absorbing dyes composed of carbazole–naphthoquinone CT complex dyes. Their results are summarized in Table II. These dyes generally show a broad absorption band and exhibit two λ_{max} in the visible and infrared regions. The λ_{max} are very sensitive to the nature of the donor/acceptor combination, and bathochromic shifts are produced depending on the strength of the donor and the acceptor. An

Table II. Values of λ_{max} for Some Intermolecular CT Complex Dyes[a]

	(13)		(14)
a	X = H	a	Y = CN, Z = NO₂
b	X = CH=N − N(Ph)₂	b	Y = CN, Z = H
		c	Y = Cl, Z = NO₂

	Donor		
	13a	13b	
Acceptor	λ (nm)[b]	λ_1 (nm)[b]	λ_2 (nm)[b]
14a	580–770	1100	604
14b	560–700	910	546
14c	535	700	490
TCNE[c]	—	950	554

[a] Ref. 10.
[b] Measured in dichloromethane.
[c] TCNE: tetracyanoethylene.

X-ray study verified the 1:1 composition of these CT complexes. There are many possible combinations of various kinds of donors and acceptors that can give infrared absorbing intermolecular CT complex dyes.

REFERENCES

1. K. A. Bello and J. Griffiths, *J. C. S. Chem. Commun.*, 1639 (1986).
2. J. Fabian and H. Hartmann, *Light Absorption of Organic Colorants*, p. 61, Springer-Verlag, Berlin (1980).
3. S. Yasui, M. Matsuoka, M. Takao, and T. Kitao, *J. Soc. Dyers Colourists 104*, 284 (1988).
4. G. A. Lezenko and A. J. Il'chenko, *Ukr. Khim. Zh. 43*, 716 (1977).
5. K. Takahashi, T. Sakae, and K. Takase, *Chem. Lett.*, 237 (1978).
6. F. A. Neugebauer, R. Bernhardt, and H. Fischer, *Chem. Ber. 110*, 2254 (1977).
7. M. Matsuoka, L. Han, H. Oka, and T. Kitao, *Chem. Express 3*, 491 (1988).
8. M. Matsuoka, T. Yodoshi, L. Han, and T. Kitao, *Dyes Pigments 9*, 343 (1988).
9. M. Matsuoka, L. Han, T. Kitao, S. Mochizuki, and K. Nakatsu, *Chem. Lett.*, 905 (1988).

II

Applications of Infrared Absorbing Dyes

9

Semiconductor Lasers

MASAYASU UENO and TONAO YUASA

I. INTRODUCTION

Various laser devices are produced as optical light sources in the optoelectronic field. Especially, semiconductor lasers[1,2] are key devices of great importance due to their small size, high efficiency, and high speed for direct modulation.

Intensive research and development efforts have been made on semiconductor lasers, since continuous-wave (CW) operation at room temperature was first reported in 1970.[3] The goals of laser device development include high reliability, which is related to.high-purity crystal growth, optimization of laser structure, which provides stable fundamental transverse-mode operation, and high-yield reproducible processes. Much effort in technology development to produce excellent semiconductor lasers has been made, and nowadays the semiconductor lasers, especially AlGaAs/GaAs lasers[1,2] with an emission wavelength of 0.78–0.85 μm and InGaAsP/InP lasers[4] with an emission wavelength of 1.1–1.6 μm, are produced commercially by various companies.

Short-wavelength lasers of AlGaAs/GaAs are mainly used as light sources in a variety of information processing systems, such as compact-disk players, laser-disk players, optical-disk drivers, and laser printers. On the other hand, long-wavelength lasers of InGaAsP/InP are mainly used as light sources[5] in optical communication systems.

MASAYASU UENO and TONAO YUASA ● Opto-Electronics Research Laboratory, NEC Corporation, Miyamae-ku, Kawasaki, Kanagawa 213, Japan.

In this chapter, the present status of semiconductor lasers, mainly AlGaAs/GaAs and InGaAsP/InP lasers, is described. In Section II, the theory of lasing operation and lasing characteristics are summarized. In the history of laser development, semiconductor lasers were improved from homojunction and single-heterojunction lasers to double-heterojunction (DH) lasers which realize CW operation at room temperature. Here, DH lasers are described.

In Section III, transverse-mode-controlled lasers are described in detail, and the reliability of AlGaAs/GaAs and InGaAsP/InP lasers is discussed. The longitudinal mode lasers, that are required for high-bit-rate, long-distance optical communication systems, are also described. In Section IV, the lasing properties that are required for lasers in various applications are described. Lasing wavelength ranges, high output power, and noise properties of semiconductor lasers are discussed. New trends in the development of semiconductor lasers are discussed in the latter part of this chapter.

II. LASING CHARACTERISTICS

This section is concerned with the fundamental structure and operating characteristics of double-heterojunction semiconductor lasers. Fundamental aspects of stimulated emission in semiconductor lasers are discussed. The recent developments in the fabrication technology to produce both AlGaAs/GaAs lasers and InGaAsP/InP lasers are also mentioned. After that, fundamental lasing characteristics are briefly summarized, with emphasis on optical output power versus current characteristics, mode properties, and astigmatism.

A. Fundamental Structure

In applications, semiconductor lasers are usually sealed in a package, and the semiconductor laser chip is bonded to the heat sink to suppress temperature rise, which increases threshold currents and decreases the operation life of the semiconductor lasers.[1,2] The laser beam is emitted through a transparent glass window which is placed in front of the laser chip.

Figure 1 shows a schematic representation of a double-heterojunction semiconductor laser and the emission mode structures which are built into the laser. As shown in Figure 1, the layer structure in a double-heterojunction semiconductor laser consists basically of the active layer, which is made of direct bandgap semiconductor compounds, and the double-heterojunction

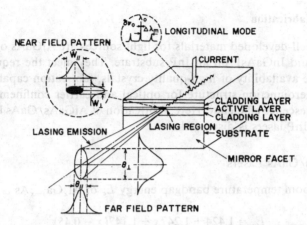

Figure 1. Transverse and longitudinal modes in a double-heterojunction laser.

structure, in which the active layer is sandwiched between higher-bandgap *p*-type and *n*-type cladding layers.

In the direct bandgap active layer, the minimum in the conduction band occurs at the same point in the Brillouin zone as the maximum in the valence band, so that stimulated emission occurs directly by recombination of electrons and holes, which are injected into the active layer from the *n*-type cladding layer and the *p*-type cladding layer, respectively. The double heterostructure provides both a potential barrier and dielectric waveguide that confine the injected carriers and the optical emission into the active layer.[1,2]

The lasing operation is attained by providing feedback of the optical emission. The laser oscillation is usually produced by use of cleaved facets for a medium with optical gain to form a Fabry–Perot resonator along the cavity axis.

Lasing begins when the total cavity loses are overcome by the optical gain. The relationship between active layer gain *g* of threshold, at which lasing begins, and cavity losses is given by[2]

$$\Gamma g = a_{CL}(1 - \Gamma) + \Gamma a_{fc} + \frac{1}{2L} \ln\left(\frac{1}{R_1 R_2}\right)$$

where Γ is the confinement factor which represents the fraction of the optical mode energy propagation in the active layer, a_{CL} is the absorption coefficient of the cladding layers at the same emission wavelength, a_{fc} is the free-carrier absorption loss in the active layer, L is cavity length, and R_1 and R_2 are the refractivities of the two mirror facets.

B. Laser Fabrication

The well-developed materials for light sources are AlGaAs on a GaAs substrate and InGaAsP on an InP substrate. They meet the requirements such as the availability of high-quality crystals, fabrication capability of a double-heterojunction structure for optical and carrier confinements, and high luminescence efficiency. Here, fabrication of AlGaAs/GaAs lasers and InGaAsP/InP lasers is discussed.

1. AlGaAs/GaAs Lasers

The room temperature bandgap energy E_g of $Al_xGa_{1-x}As$ is given by[2]

$$E_g = 1.424 + 1.247x + 1.147(x - 0.45)^2$$

where x is the Al mole fraction. The laser emission energy is about 0.03 eV less than the bandgap energy of the active layer because of bandgap shrinkage[2,6] at threshold. For a typical laser with an emission wavelength of 0.78 μm, the Al mole fraction x in the active layer is 0.15 and the active layer thickness is $0.04 \sim 0.05$ μm, while x in the cladding layers is $0.45 \sim 0.50$ and their thickness is $1.0 \sim 2.0$ μm.

AlGaAs/GaAs lasers have been usually fabricated by liquid-phase epitaxy using a sliding boat.[1,2] Recently, new fabrication methods such as molecular-beam epitaxy (MBE)[7] and metal–organic vapor-phase epitaxy (MOVPE)[8] have been widely used because of their capability of providing better uniformity and thickness control.

2. InGaAsP/InP Lasers

The room temperature bandgap energy E_g of $In_{1-x}Ga_xAs_yP_{1-y}$ is approximately given by[4,9]

$$E_g = 1.35 - 0.72y + 0.12y^2$$

where y is the As mole fraction. In the case of InGaAsP material, the lattice matching condition must be considered. This is given by[4]

$$x = \frac{0.4526y}{1.0 - 0.031y}$$

where x is the Ga mole fraction. The laser emission energy is about 0.03 eV less than the bandgap energy. For typical lasers with emission wavelengths of 1.3 and 1.55 μm, where the absorption loss becomes lowest in the optical fibers,[10] the active layer is $In_{0.72}Ga_{0.28}As_{0.61}P_{0.39}$ and $In_{0.58}Ga_{0.42}As_{0.90}P_{0.10}$, respectively, and the active layer thicknesses are about 0.2 μm, while the cladding layers are InP and their thicknesses are 1.0 ~ 2.0 μm.

InGaAsP/InP lasers have been usually fabricated by liquid-phase epitaxy using a sliding boat.[11] The lattice mismatching can be kept to less than 0.03% by controlling melt composition and growth temperature. Recently, VPE[12] and MOVPE[13] have been widely used for crystal growth. Both methods do not have the problem of meltback of the active layer, which often occurs in liquid-phase epitaxial growth of lasers with emission wavelengths longer than 1.5 μm.

C. Lasing Characteristics

1. Optical Output Power versus Current Characteristics

The main key parameters on the optical output power versus current characteristics are the threshold current I_{th}, which depends on the diode area and the threshold current density J_{th}, and the differential quantum efficiency η_d. Both I_{th} and η_d depend on the initial device structure and vary with temperature.

The temperature dependence of the threshold current densities of AlGaAs/GaAs lasers and InGaAsP/InP lasers is approximately expressed by

$$J_{th} = J_0 \exp(\Delta T / T_0)$$

where J_0 is the pulsed threshold current density at room temperature and ΔT is the temperature rise at the light-emitting region from room temperature. The characteristic temperature T_0 is typically 120–150 K in AlGaAs/GaAs lasers,[14] while T_0 is typically 60–80 K in InGaAsP/InP lasers.[15] Typical examples of the temperature dependence of the threshold current in AlGaAs/GaAs and InGaAsP/InP lasers are shown in Figure 2a and 2b, respectively.

The temperature dependence of the threshold current is largely the result of exciting greater numbers of carriers above the recombination energy levels as the temperature increases.[2] Another significant contribution to the temperature dependence of threshold current in AlGaAs/GaAs lasers can be the leakage of electrons from the active region into the p-type cladding layer.[16]

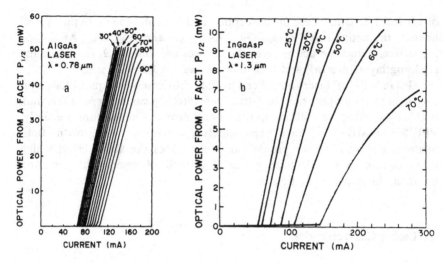

Figure 2. CW output power versus current characteristics for AlGaAs(a) and InGaAsP(b) lasers with changing the ambient temperature.

In the case of InGaAsP/InP lasers, in addition to the mechanisms described above, recombination through the Auger process makes a significant contribution,[17] resulting in small T_0.

2. Mode Pattern

At lasing, three kinds of standing wave are built up in a laser cavity as shown in Figure 1.[1,2] One is longitudinal modes, which are related to the cavity length, and shows spectral modes. The other two standing waves form the transverse modes, which are built up parallel and perpendicular to the junction plane, respectively. The transverse mode perpendicular to the junction plane is formed in the dielectric waveguide which is made by the active layer and two cladding layers as shown in Figure 1. The transverse mode parallel to the junction plane (lateral transverse mode) depends on the preparation of the side wall structure along the junction plane as discussed in Section III. The intensity distributions of transverse modes on the facet are called the near-field pattern. On the other hand, the transverse-mode distributions which are measured far from the facet are called the far-field pattern. The far-field pattern is the Fourier integral of the near-field distribution.[2]

3. Astigmatism

In applications of the laser beam in the optoelectronic field, astigmatism[18] is a big problem, because it is difficult to collimate the laser beam when astigmatism is large. The astigmatism usually occurs because the beam waist parallel to the junction is virtually within the laser device, while the beam waist perpendicular to the junction plane is usually at the mirror facet. The distance between the position of the beam waist parallel and perpendicular to the junction plane, which is defined as astigmatism, is usually $20 \sim 30 \mu$m for gain-guided lasers and $2 \sim 5 \mu$m for index-guided lasers. These lasers are described in Section III. Laser structure control is necessary to obtain small astigmatism.

III. LASER DESIGN

The lasing-mode stabilization, especially lateral transverse mode stabilization, is one of the most important to practical application of lasers because lateral transverse mode instability causes so-called kinks in light output power versus current characteristics, which are accompanied by anomalous lasing behaviors, such as excess intensity fluctuation, beam direction shifts, and deterioration of modulation characteristics.[19,20] The theoretical analysis about kinks mode by R. Lang[21] indicate that spatial hole-burning, the negative dependence of refractive index on the carrier density, and the lack of complete symmetry in the lateral direction in any real laser structures are the three critical factors responsible for the lateral transverse mode instability. After intensive efforts, various laser cavity structures have been introduced to stabilize the laser transverse mode. In this section, transverse-mode-stabilized lasers are presented.

Reliability is the most important issue for the use of lasers in various applications. The reliability of both AlGaAs/GaAs lasers and InGaAsP/InP lasers is discussed in the second part of this section.

Stable single-longitudinal-mode operation is ideal for high-bit-rate, long-distance optical communications. Thus, longitudinal-mode control is well developed on long-wavelength InGaAsP/InP lasers. In the latter part of this section, this longitudinal-mode control is also discussed.

A. Mode Stabilization

In order to obtain high-quality lasers, with such features as low threshold current, high efficiency, and stabilized lateral transverse mode, various laser structures have been successfully demonstrated.[22]

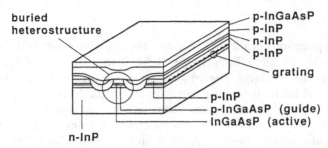

Figure 3. Structure diagram of DFB-DC-PBH laser with first-order grating. Both facets were cleaved.

There are two basic approaches to obtain these high-quality lasers. One method is introduction of rigid-waveguide structure, and the other method is narrowing the current injection path. Lasers with a transverse mode guided by a rigid waveguide are called index-guided lasers,[22] and those with a narrow current path are called gain-guided lasers.[22]

A typical structure of index-guided lasers is that of the buried heterostructure (BH) laser,[13,24] as shown in Figure 3. In this laser, a mesa-shaped double heterostructure is buried with low-refractive-index material by a second epitaxial growth. Then the active region is completely embedded by cladding regions with refractive index lower than that of the active region.

In the second method for realizing index-guided lasers, an effective refractive index difference is established along the active layer in accordance with the waveguide layer thickness variation.[25,26] The third method of index-guide is to grow a double-heterostructure layer adjacent to a grooved substrate.[27,28] In this case, the index-guide is realized by a complex index step originating from the optical loss variations at the substrate.

In contrast to the index-guided lasers, no rigid waveguide is formed around the active laser in gain-guided lasers.[29] In order to eliminate the mode instability, the injected current path is made sufficiently narrow.[30,31,32]

B. Reliability

In AlGaAs/GaAs lasers with emission wavelengths ranging from 0.78 to 0.85 μm and InGaAsP/InP lasers with emission wavelengths ranging from 1.1 to 1.6 μm, not only transverse-mode stabilization but also reliability has been achieved.

1. AlGaAs/GaAs Lasers

AlGaAs/GaAs laser degradation can be separated into three categories, which are "dark-line-defect" (DLD) formation, facet damage, and internal gradual degradation.[2]

The DLD[33] is a network of dislocations[34] that causes rapid degradation within hours after the start of CW operation. The long-term efforts, which include the care paid to cleanliness and gas purity during heteroepitaxial wafer growth, eliminate DLD formation in fabricated lasers.

Facet damage is divided into catastrophic optical damage and facet erosion. Catastrophic optical damage,[35,36] which depends on output power intensity density, can be controlled by proper limitation of the power emission. Facet erosion, which results from oxidation of the mirror facet at the active region, generally occurs over long periods of lasing time.[37] Facet erosion is prevented by coating the facet with a dielectric film, such as Al_2O_3, Si_3N_4, SiO_2 or their combinations.[38] Internal gradual degradation remains the limiting factor on laser life. Current estimates of room temperature life by extrapolation indicate a median time to failure (MTTF) of between 10^5 and 10^6 hours.[39]

2. InGaAsP/InP Lasers

The facets of InGaAsP/InP lasers are much stronger than those of AlGaAs/GaAs lasers.[40] Thus, the catastrophic failure limit in InGaAsP/InP lasers is extremely high, and usually the output power level is limited by temperature rise before it reaches the catastrophic failure power level. Moreover, slow degradation due to oxidation is also hardly observed without facet coating. Room temperature life is estimated as much longer than that of AlGaAs/GaAs lasers.[41]

C. Longitudinal-Mode Control

Intentional control of the laser longitudinal mode is basically possible by grating feedback. Distributed feedback (DFB) lasers[42] and distributed Bragg reflector (DBR) lasers[42] have been investigated mainly in the development of InGaAsP/InP laser technology because stable single-longitudinal-mode operation is a necessary laser property for long-distance optical communication systems.

The grating period V is given by

$$V = M\lambda/(2.0N_g)$$

where λ is lasing wavelength, N_g is the effective waveguide refractive index, and M is an integer. In the case of InGaAsP/InP lasers, $V = 0.24~\mu$m ($M = 1$) when $\lambda = 1.55~\mu$m and $N_g = 3.23$, while $V = 0.20~\mu$m ($M = 1$) when $\lambda = 1.3~\mu$m and $N_g = 3.24$. Recently, $\lambda/4$ shift DFB lasers have been fabricated to provide high speed and high efficiency dynamic single longitudinal mode light sources.[43] Except for the single-longitudinal-mode property, the other lasing characteristics and the reliability of 1.3- and

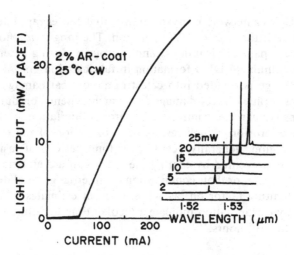

Figure 4. CW single-longitudinal-mode operation at room temperature for an antireflectively coated DFB laser.

1.55-μm-band DFB lasers are comparable to those of Fabry–Perot lasers. Nowadays, DFB lasers with emission wavelengths of both 1.3 and 1.55 μm are produced commercially.

As a typical example, Figure 4 depicts light output power versus current characteristics under CW and single-longitudinal-mode operation at various CW output levels, measured with a 1.55-μm-band antireflectively (AR) coated DFB double-channel planar buried heterostructure (DC-PBH) laser[44] which structure is shown in Figure 3.

IV. OTHER REQUIRED LASING PROPERTIES

The lasing properties that are required for lasers in various applications are described in this section. First, lasers with different emission wavelengths from those of AlGaAs/GaAs and InGaAsP/InP lasers are introduced. These lasers are attractive light sources for an increasing range of applications. Emission wavelengths are divided into visible wavelengths and longer wavelengths. Visible lasers with emission wavelengths shorter than the 0.78-μm wavelength of AlGaAs/GaAs lasers have been investigated as light sources mainly in optical information processing systems. Especially, the technology for the fabrication of AlGaInP/GaAs lasers with emission wavelengths shorter than ~0.68 μm has progressed rapidly in recent years, and reliability has also been established. Longer-wavelength lasers with emission wavelengths longer than ~1.6 μm have also been investigated as light sources for future communication systems.

Second, high-power lasers, especially fundamental-mode AlGaAs/GaAs lasers, are discussed. The demand for these lasers is increasing for light sources in erasable direct read-after-write (EDRAW) optical disk systems. Some structures which have been successfully introduced as high-power lasers are presented.

Finally, the noise properties of semiconductor lasers are discussed. Decreasing noise effects is necessary to apply these lasers as light sources in optical systems, such as optical disk memory systems and optical fiber communication systems. Several ways for decreasing noise which are used mainly in AlGaAs/GaAs lasers are mentioned here.

A. Lasers with Various Emission Wavelengths

1. Visible-Light Lasers

Visible-light lasers with emission wavelengths shorter than the 0.7-μm-wavelength band have been intensively studied. Visible-light lasers are attractive light sources for the following reasons. The shorter the emission wavelength becomes, the smaller is the spot size that is obtained. Accordingly, the recording data density on an optical disk can be increased remarkably. Moreover, visible-light lasers increase range of applications, such as in optical information processing systems which presently use He–Ne lasers.

Figure 5 shows emission wavelengths for various III–V mixed crystals, which include some candidates for use in visible light lasers. In particular, AlGaInP/GaAs visible-light lasers with emission at 0.67-0.68 μm can be stably operated for more than 2000 hours at room temperature,[45] and these lasers are now begining to be produced commercially. It has also been reported that AlGaInP/GaAs lasers have an emission wavelength of 0.62 μm at 3°C.[46] This wavelength is shorter than that of He–Ne lasers.

ZnSSeTe/GaAs lasers have many problems that remain to be solved. However, these lasers are anticipated as blue-green lasers.

2. Longer-Wavelength Lasers

Semiconductor lasers with emission wavelengths exceeding ~2 μm have also been studied for many years.[47] The useful applications of these lasers include applications to high-resolution gas spectroscopy and air pollution monitoring. Most recently, much attention has been paid to

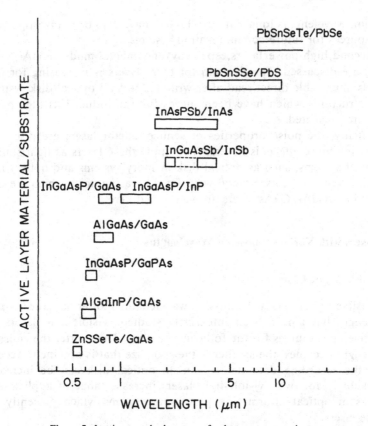

Figure 5. Lattice-matched systems for heterostructure lasers.

ultra-low-loss optical fiber materials in the near- and middle-infrared regions, such as thallium bromide and chalcogenide glass materials.[48] These lasers are expected to be used in the near future as light sources in fiber communication systems.

Figure 5 shows expected emission wavelengths for various III–V and II–VI mixed crystals which give lasers with wavelengths exceeding ~2 μm.[47] Small energy gap of longer wavelength laser produces large band-to-band Auger process, that decreases the minority-carrier lifetime, and large free-carrier absorption. Then, optical gain required for laser oscillation increases with increasing emission wavelength. Until now, room temperature laser operation has been reported on AlGaAs/GaSb[49] and InGaAsSb/AlGaAsSb[50] DH lasers.

B. High-Power Fundamental-Mode AlGaAs/GaAs Lasers

There is an every-increasing demand for high-power fundamental-mode DH lasers to be used in applications, such as optical recording, high-speed laser printing, and space communications.

The maximum optical power available from an AlGaAs/GaAs DH laser is limited by the catastrophic optical damage (COD)[35] which occurs typically when the optical power density exceeds a critical value of 5–6 MW cm^{-2} in pulsed operation and ~1 MW cm^{-2} (3–4 mW μm^{-1}) in CW operation.[35,51] It was shown by Henry *et al.*[52] that COD results from local melting absorption in depleted regions, which have effectively lower band-gap, adjacent to mirror surfaces.

Moreover, CW optical power is usually saturated below the COD threshold, because the CW optical power is limited by the temperature rise in the light-emitting region, which depends on pulsed lasing characteristics and thermal resistance.

In order to increase fundamental-mode power capability, two basic approaches have been taken.

The first method involves increasing the lasing spot size both perpendicular and parallel to the junction plane. In this case, it is also necessary to introduce a mode-dependent loss mechanism in order to discriminate against higher-order mode oscillation.

The easier step of this first method is increasing the lasing spot size perpendicular to the junction plane. There are two ways of achieving this. One way is thinning the active layer in the DH structure.[53] As a typical example, Figure 6 shows the light output power versus current characteristics and far-field intensity profiles parallel to the junction plane at various CW output power levels, that are measured for an index-guided AlGaAs/GaAs laser with a thin active layer thickness (0.04 μm), and 10% and 75% facet coating, respectively (the laser structure is called a plano-convex waveguide (PCW)[25]). Because of thin active layer and low-reflection, such as 10%, coated front facet, the maximum CW power, limited by COD, is as high as 150 mW, and fundamental transverse-mode operation has been achieved at up to 100 mW. Thinning the active layer just adjacent to mirror surfaces is also used for increasing COD level. This method is more effective because the fundamental-mode CW power output increases without an increase in the threshold current.[54]

The other way to increase spot size is to create a large optical cavity (LOC) structure, which involves the growth of an additional guide layer with an index intermediate between that of the active layer and that of the cladding layer.[55,56]

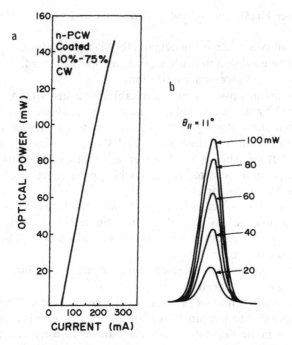

Figure 6. Lasing characteristics of a typical new-plano-convex-waveguide (n-PCW) laser: (a) CW output power versus current characteristics; (b) far-field patterns parallel to the junction plane at various CW output power levels.

The second method for increasing output power is to eliminate facet degradation by making window structures with effectively higher bandgap regions near mirrors. A typical example is the window stripe (WS) laser.[57,58] The WS laser has heavily doped n-type window regions near mirrors and a Zn-diffused overcompensated p-type light-emitting region in the central part. The absorption edge in the heavily doped n-type window regions shifts to higher energy due to Burstein shift,[59] while the effective bandgap in the Zn-diffused overcompensated light-emitting region shrinks due to band tailing.[60] Thus, laser light absorption near the mirror surfaces is markedly reduced, resulting in high-power operations.

C. Low-Noise Lasers

Noise affects the performance of a system using a laser diode as a light source, because the intensity of the laser output fluctuates. In particular, noise degrades the quality of an optical disk and error rate performance in optical fiber communication systems.[61,62] The noise in semiconductor lasers has several causes.

The most intrinsic mechanism of noise is the random nature of carrier recombination, which gives the quantum noise. The theoretical analysis by McCumber[63] and Haug[64] showed that the quantum noise intensity is maximum at laser threshold.

Another mechanism of noise is the mode competitions that take place when almost singular longitudinal mode hops toward a neighboring mode as the ambient temperature or device current changes.[65,66] The mode hopping causes significant noise at a frequency below several tens of megahertz. The mode-hopping noise can be eliminated by longitudinal-mode stabilization, such as distributed feedback (DFB) structure. It can also be eliminated by preventing mode competition during mode hopping. It has been reported that Te in an n-type AlGaAs cladding layer stabilizes the longitudinal mode in AlGaAs/GaAs lasers,[67] and decreases the noise level.

Laser oscillation becomes unstable when feedback of the laser light by external reflection occurs.[66] In this case, the interactions created between the laser cavity modes (longitudinal modes in the laser cavity) and the external cavity modes, which are formed between a laser facet and an external reflector, changes emission spectra considerably,[68] resulting in feedback noise. Even at small feedback rate, feedback noise has been observed. The feedback noise level can be higher than the other noises and is consequently of more concern. Various methods have been investigated to eliminate feedback noise.

One method of eliminating feedback noise is to decrease the coherency of the longitudinal mode, which is most sensitive to the feedback light. The effective ways to achieve this are by high-frequency modulation of the lasers[66] or by making self-pulsations.[69] In both cases, the linewidth of individual longitudinal modes broadens and coherency decreases.

Another method to suppress feedback noise involves coating high reflective films on high mirror facets in the lasers. In these lasers, spontaneous emission in the active layer increases, and it suppresses the nonlasing spectrum. Then, one main longitudinal mode stabilizes. In addition, reflective light penetrating into the laser cavity decreases because of the high reflective film on the facets.

As a typical example, Figure 7 shows relative intensity noise (RIN) as a function of the optical feedback ratio for a high-power AlGaAs/GaAs laser emitting at 0.78 μm, with a 20% coated front facet and a 75% coated rear facet, under high-frequency modulation at 3 mW optical output. The RIN value is as low as -126 dB/Hz up to 7% feedback ratio.[70] These results indicate that high-frequency modulation and high reflective coating are effective in decreasing noise in semiconductor lasers.

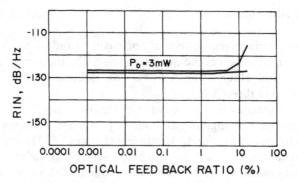

Figure 7. Relative intensity noise (RIN) versus optical feedback ratio to AlGaAs/GaAs laser. Output power from a facet is 3 mW.

V. NEW DEVELOPMENTS

In this section, new trends in research on semiconductor lasers are discussed. Especially, this section is concerned with the newly developed quantum well heterojunction lasers and optoelectronic integrated circuits (OEIC).

A. Quantum Well Heterojunction Lasers

Newly developed growth methods such as MBE and MOVPE allow very low growth rate, ultra-thin-layer growth (<10 Å), and good control for desired abrupt change. These features make it possible to create quantum well (QW) heterojunction lasers.[71] In these heterostructures, free motion of the carriers occurs only in the two directions perpendicular to the growth direction, the motion in the third direction, namely, parallel to the growth direction, being restricted to a well-defined space. The two-dimensional nature of electron motion in the QW heterostructure produces several unique features in semiconductor lasers. These quantum-size effects shorten the emission wavelength due to the radiative transition between confined states and significantly reduce the threshold current density and its temperature dependence as a result of the modification in the density of state function of the electrons.

Two kinds of lasers based on these QW heterostructure effects have been investigated. One is multiple-stripe monolithic laser arrays as light sources of extremely high power. Optical power in excess of 134 W from one facet has been obtained for a pulse width of 150 μs (quasi-CW operation) from monolithic AlGaAs/GaAs laser arrays with 1-cm emitting widths, in which the active region consists of four quantum wells, each about 80 to 10 Å thick, separated by three barriers, each about 40 Å thick.[72]

The other type of laser in which QW heterostructure effects are utilized is ultra-low-threshold lasers. It is reported that the buried heterostructure type graded index separate confinement heterostructure single QW laser (BH GRIN SCH SQW), in which ~80% reflectivities of both facets are provided, obtained 0.9 mA CW threshold current.[73]

QW lasers are also very favorable as light sources for optical disk memory systems because of their low noise characteristics. It is reported that the laser operates at 50 mW and exhibits low feedback noise (less than −135 dB/Hz at 3 mW).[74]

B. Optoelectronic Integrated Circuits

Recently, optoelectronic integrated circuits (OEIC)[75] have been widely investigated. The monolithic integration of optical and electrical com-

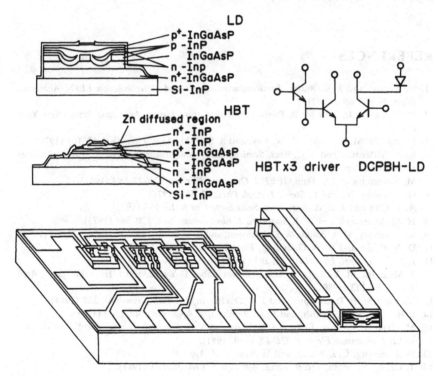

Figure 8. Laser diode (LD)–heterobipolar transistors (HBTs) transmitter optoelectronic integrated circuit (OEIC).

ponents is a natural trend of semiconductor technology. The advantages of monolithic integration are recognized as an improvement in compactness and reliability, a reduction in cost, and an improvement in performance. Such potential advantages have led OEIC research in two major directions.

One research direction is to improve the performance of optoelectronic devices for use in high-speed optical fiber communication systems by monolithic integration of electronic circuits.

The other research direction is to improve electronic circuit performance by introducing optical signal interconnection. Optical signal interconnection has such advantages as freedom from the strong capacitance problem in electrical interconnection wires, electrical isolation, and small size.

As a typical example, Figure 8 shows a transmitter OEIC consisting of a laser diode and a driving circuit.[76]

Future optical systems may be replaced by small OEIC chips. For this purpose, a large amount of research effort has been undertaken to realize high-performance OEIC chips.

REFERENCES

1. H. Kressel and J. K. Butler, *Semiconductor Lasers and Heterojunction LEDs*, Academic Press, New York (1977).
2. H. C. Casey, Jr., and M. B. Panish, *Heterostructure Lasers*, Academic Press, New York (1978).
3. I. Hayashi, M. B. Panish, P. W. Foy, and S. Sumski, *Appl. Phys. Lett. 17*, 109 (1970).
4. R. K. Willardson and A. C. Beer, *Semiconductors and Semimetals, vol. 22*, Part C, Chapter 1, Academic Press (1985).
5. M. Nakamura and S. Tsuji, *IEEE J. Quantum Electron. QE-17*, 994 (1981).
6. H. C. Casey, Jr., and F. Stern, *J. Appl. Phys. 47*, 631 (1976).
7. A. Y. Cho and J. R. Arthur, *Prog. Solid State Chem. 10*, 157 (1975).
8. H. M. Manasevit and W. I. Simpson, *J. Electrochem. Soc. 120*, 569 (1973).
9. R. E. Nahory, M. A. Pollak, W. D. Johnston, and R. L. Burns, *Appl. Phys. Lett. 33*, 659 (1978).
10. D. N. Payne and W. A. Gambling, *Electron. Lett. 11*, 176 (1975).
11. J. J. Hsieh, *Appl. Phys. Lett. 28*, 283 (1976).
12. T. Mizutani, M. Yoshida, A. Usui, H. Watanabe, T. Yuasa, and I. Hayashi, *Jpn. J. Appl. Phys. 19*, L113 (1980).
13. M. Razeghi, B. Decremoux, and J. P. Duchemin, *J. Cryst. Growth 68*, 389 (1984).
14. I. Hayashi, M. B. Panish, and F. K. Reinhart, *J. Appl. Phys. 42*, 1929 (1971).
15. M. Ueno, I. Sakuma, T. Furuse, Y. Matsumoto, H. Kawano, Y. Ide, and S. Matsumoto, *IEEE J. Quantum Electron. QE-17*, 1930 (1981).
16. M. Ettenberg, C. J. Nuese, and H. Kressel, *J. Appl. Phys. 50*, 2949 (1979).
17. T. Uji, K. Iwamoto, and R. Lang, *Appl. Phys. Lett. 38*, 193 (1981).
18. D. D. Cook and F. R. Nash, *J. Appl. Phys. 46*, 1660 (1975).
19. T. L. Paoli, *IEEE J. Quantum Electron. QE-12*, 770 (1976).

20. K. Kobayashi, R. Lang, H. Yonezu, I. Sakuma, and I. Hayashi, *Jpn. J. Appl. Phys. 16*, 207 (1977).
21. R. Lang, *IEEE J. Quantum Electron. QE-15*, 718 (1979).
22. R. K. Willardson and A. C. Beer, *Semiconductor and Semimetals, vol. 22*, Part C, Chapter 2. Academic Press (1985).
23. T. Tsukada, *J. Appl. Phys. 45*, 4899 (1974).
24. I. Mito, M. Kitamura, K. Kobayashi, and K. Kobayashi, *Electron. Lett. 18*, 953 (1982).
25. Y. Ide, T. Furuse, I. Sakuma, and K. Nishida, *Appl. Phys. Lett. 36*, 121 (1980).
26. M. Ueno, R. Lang, S. Matsumoto, H. Kawano, T. Furuse, and I. Sakuma, *IEE Proc. 129*, 218 (1982).
27. K. Aiki, M. Nakamura, T. Kuroda, J. Umeda, R. Ito, N. Chinone, and M. Maeda, *IEEE J. Quantum Electron. QE-14*, 89 (1978).
28. T. Hayakawa, N. Miyauchi, S. Yamamoto, H. Hayashi, S. Yano, and T. Hijikata, *J. Appl. Phys. 53*, 7224 (1982).
29. H. Yanezu, I. Sakuma, K. Kobayashi, T. Kamejima, M. Ueno, and Y. Nannichi, *Jpn. J. Appl. Phys. 12*, 1585 (1973).
30. T. Kobayashi, H. Kawaguchi, and Y. Furukawa, *Jpn. J. Appl. Phys. 16*, 601 (1977).
31. M. Ueno, *Jpn. J. Appl. Phys. 16*, 1399 (1977).
32. P. Marshall, E. Schlosser, and C. Wölk, *Electron. Lett. 15*, 38 (1979).
33. H. Yonezu, I. Sakuma, T. Kamejima, M. Ueno, K. Nishida, Y. Nannichi, and I. Hayashi, *Appl. Phys. Lett. 24*, 18 (1974).
34. P. M. Petroff and R. L. Hartmann, *Appl. Phys. Lett. 23*, 469 (1973).
35. B. W. Hakki and F. R. Nash, *J. Appl. Phys. 45*, 3907 (1974).
36. P. A. Kirkby and G. H. B. Thompson, *Appl. Phys. Lett. 22*, 638 (1973).
37. T. Yuasa, M. Ogawa, K. Endo, and H. Yonezu, *Appl. Phys. Lett. 32*, 119 (1978).
38. T. Yuasa, K. Endo, T. Torikai, and H. Yonezu, *Appl. Phys. Lett. 34*, 685 (1979).
39. M. Ettenberg and H. Kressel, *IEEE J. Quantum Electron. QE-16*, 186 (1980).
40. M. Fukuda, K. Takahei, G. Iwane, and T. Ikegami, *Appl. Phys. Lett. 41*, 18 (1982).
41. K. Mizuishi, *IEEE J. Quantum Electron QE-19*, 457 (1983).
42. R. K. Willardson and A. C. Beer, *Semiconductors and Semimetals, vol. 22*, Part B, Chapter 4, Academic Press (1985).
43. M. Yamaguchi, S. Takano, S. Fujita, I. Cha, and I. Mito, *Optoelectronics 3*, 257 (1988).
44. M. Kitamura, M. Yamaguchi, S. Murata, I. Mito, and K. Kobayashi, *IEEE J. Lightwave Technol. T-2*, 363 (1984).
45. A. Gomyo, K. Kobayashi, S. Kawata, I. Hino, and T. Suzuki, *Electron. Lett. 23*, 85 (1987).
46. K. Kobayashi, I. Hino, and T. Suzuki, *Appl. Phys. Lett. 46*, 7 (1985).
47. R. K. Willrdson and A. C. Beer, *Semiconductors and Semimetals, vol. 22*, Part C, Chapter 3, Academic Press (1985).
48. K. Shinohara, T. Akamatsu, and R. Ueda, *Jpn. J. Appl. Phys. 20*, 439 (1981).
49. L. M. Dolginov, A. E. Drakin, L. V. Druzhinita, P. G. Eliseev, M. G. Milvidsky, V. A. Skripken, and B. N. Sverdlov, *IEEE J. Quantum Electron. QE-17*, 593 (1981).
50. N. Kobayashi and Y. Horikoshi, *Jpn. J. Appl. Phys. 19*, L641 (1980).
51. H. Kressel and I. Ladany, *RCA Rev. 36*, 230 (1975).
52. C. H. Henry, P. M. Petroff, R. A. Logan, and F. R. Merritt, *J. Appl. Phys. 50*, 3721 (1979).
53. M. Wada, K. Hamada, H. Shimizu, T. Sugino, F. Tajiri, K. Itoh, G. Kano, and I. Teramoto, *Appl. Phys. Lett. 42*, 853 (1983).
54. T. Shibutani, M. Kume, K. Hamada, H. Shimizu, K. Itoh, G. Kano, and I. Teramoto, *IEEE J. Quantum Electron. QE-23*, 760 (1987).
55. H. F. Lockwood, H. Kressel, H. S. Sommers, Jr., and F. Z. Hawrylc, *Appl. Phys. Lett. 17*, 499 (1970).

56. D. Botez, *Appl. Phys. Lett.* *36*, 190 (1980).
57. H. Yonezu, I. Sakuma, T. Kamejima, M. Ueno, K. Iwamoto, I. Hino, and I. Hayashi, *Appl. Phys. Lett.* *34*, 637 (1979).
58. M. Ueno, *IEEE J. Quantum Electron.* *QE-17*, 2113 (1981).
59. H. C. Casey, Jr., D. D. Sell, and K. W. Wecht, *J. Appl. Phys.* *46*, 250 (1975).
60. P. D. Dapkus, N. Holonyak, Jr., J. A. Rossi, F. V. Williams, and D. A. High, *J. Appl. Phys.* *40*, 3300 (1969).
61. K. Peterman and G. Arnold, *IEEE J. Quantum Electron.* *QE-18*, 543 (1982).
62. R. E. Epworth, *Proc. 4th Europ. Conf. Opt. Commun.* 492 (1978).
63. D. E. McCumber, *Phys. Rev.* *141*, 306 (1966).
64. H. Haug, *Phys. Rev.* *184*, 338 (1969).
65. N. Chinone, K. Takahashi, T. Kajimura, and M. Ojima, *Proceedings of the 8th IEEE Semiconductor Laser Conference*, Ottawa, Ontario, Canada, (Sept. 25, 1982).
66. M. Ojima, A. Arimoto, N. Chinone, T. Gotoh, and K. Aiki, *Appl. Opt.* *25*, 1404 (1986).
67. J. A. Copeland, *IEEE J. Quantum Electron.* *QE-16*, 721 (1980).
68. R. Lang and K. Kobayashi, *IEEE J. Quantum Electron.* *QE-16*, 347 (1980).
69. J. P. van der Ziel, J. L. Merz, and T. L. Paoli, *J. Appl. Phys.* *50*, 4620 (1979).
70. I. Komazaki, M. Uchida, M. Nido, S. Ishikawa, K. Endo, K. Hara, and T. Yuasa, *Electron. Lett.* *25*, 294 (1989).
71. N. Holonyak, R. M. Kolbas, R. D. Dupuis, and P. D. Dapkus, *IEEE J. Quantum Electron.* *QE-16*, 170 (1980).
72. G. L. Harnagel, P. S. Cross, C. R. Lennon, M. Devito, and D. R. Scifres, *Electron. Lett.* *23*, 743 (1987).
73. P. L. Derry, H. Z. Chen, H. Morkoc, A. Yariv, K. Y. Lau, N. Bar-Chaim, K. Lee, and J. Rosenberg, *J. Vac. Sci. Technol.* *B6*, 689 (1988).
74. M. Nido, K. Endo, S. Ishikawa, M. Uchida, I. Komazaki, K. Hara, and T. Yuasa, *Electron. Lett.* *25*, 277 (1989).
75. A. Suzuki, K. Kasahara, and M. Shikada, *IEEE J. Lightwave Technol.* *LT-5*, 1479 (1987).
76. Y. Inomoto, T. Terakado, and A. Suzuki, *Technical Digest of the First Optoelectronics Conference*, A6-4, Tokyo (1986).

10

Optical Recording Systems

FUMIO MATSUI

I. INTRODUCTION

Read-only optical disks such as laser disks and compact disk videos which deliver high-quality video picture and sound as well as compact disks which deliver digital sounds are becoming popular in commercial use.

Eight years have passed since the development of optical disks capable of both recording and reproducing information. However, the magnetic recording medium remains the most popular for information recording. Nevertheless, optical recording media have great capacity and effectively satisfy the need for achieving high density in the recording of various types of information.

During the 1970s, continuous study of the optical disk for the recording of information was conducted.[1,2] Development of the optical disk was accelerated particularly after the successful development of semiconductor lasers emitting wavelengths shorter than 900 nm. This eventually paved the way for the development of smaller-size recording systems, which in turn accelerated the development of new recording media.

Initially, an optical disk composed of inorganic tellurium (Te) with carbon was offered for use in 1981.[3] Later, Te alloys and/or oxides came into use as recording media.[4,5] An optical disk composed of cyanine dyes was proposed in 1985.[6,7] Since then, many studies have been conducted to develop an organic recording medium composed of various types of near-

FUMIO MATSUI • Corporate Research and Development Laboratory, Pioneer Electronic Corporation, Tsurugashima-machi, Iruma-gun, Saitama 350-02, Japan.

infrared absorbing (IR) dyes, and a nickel-complex-stabilized cyanine dye system was commercialized in 1988.[8,21]

The recording media proposed in the 1970s were sensitive to the gas laser wavelength region. However, the organic recording media proposed in the 1980s have intense absorption in the near-infrared region of the semiconductor laser. Actually, very few dyes are known for optical recording purposes. The majority of IR dyes were developed for other uses, for example, photographic dyes, and do not have the characteristics required for optical recording. Most of the IR dyes for optical recording media are newly synthesized so that a variety of characteristics can be generated according to the requirements of the application.

Some characteristics of organic recording media are summarized below.

1. The spin coating process can be effectively applied for producing thin films of the recording layer at low cost.
2. The thin-film surface can be kept smooth to minimize noise.
3. Thermal conductivity of organic media is negligible so the pit length is easily controlled. This enables compatibility of the recording media with a variety of modulation systems.
4. Organic dye media have good fastness against oxidation compared with inorganic media and have longer lifetimes.
5. Softness of organic dye media effectively prevents cracking of the thin film.
6. Organic dye media are generally nontoxic.
7. Due to the low reflection characteristics of organic thin films, multiple layers are required.
8. Some organic dye media do not have sufficient lightfastness.
9. The majority of organic recording media do not exhibit a well-defined threshold value of the recording power.

In short, despite some undesirable properties, the IR dyes have a variety of advantages for optical recording media. The following section discusses the features of optical disks composed of IR dyes.

II. CLASSIFICATION OF THE OPTICAL MEMORY DISK

An optical memory system composed of a recording medium allows the user to record and reproduce the required information. The information to be recorded or stored can be converted into electrical signals and then into optically modulated signals. The user can record information onto the optical memory disk, and it can then be optically reproduced. The user can reproduce the recorded information by reading the difference of the reflectance on light irradiation. The optical disk drive and recording/reproducing systems are designed to record and read the information by light irradiation onto the recording medium. Laser diodes which emit near-infrared light are used as the light source.

Figure 1 shows the classification of optical disks. By function, optical disks are classified into two types: the erasable type, which allows overwriting of information many times, and the write-once type, which allows the user to record information only once onto the recording medium. The principal advantage of the optical disk is its vast recording capacity. Each optical memory has a recording capacity several hundred times that of a conventional magnetic memory. Consequently, even the write-once optical memory system can be used for a wide variety of data storage.

By recording principle, optical disks can be classified into two functional modes: the heat mode and the photon mode. The heat mode records information by means of structural alternations on the medium caused by heat on light absorption. In the photon mode, recording is caused by means of photoenergy. Optical disk systems currently in commercial use are solely of the heat-mode type.

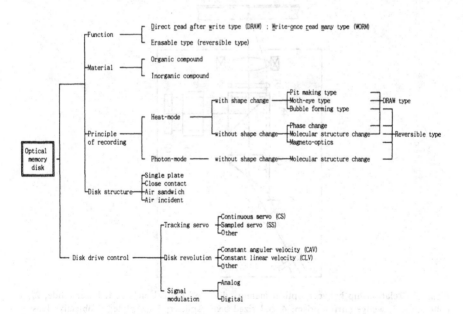

Figure 1. Classification of optical memory disks.

III. FUNCTION OF THE OPTICAL MEMORY DISK SYSTEM

Basically, the optical memory disk plays three roles: (1) output of disk-drive control signals, (2) recording information and output of the signals to reproduce the recorded information, and (3) storage of the recorded information.

As for all conventional magnetic recording media, the disk-drive system of the optical memory disk is necessary for proper recording and reproduction of information. The electrical signals delivered to the disk-drive system depend on the magnitude of the reflected light from the surface of the optical memory disk. The reflected light either increases or decreases due to interference and/or diffraction by the preformatted pits and due to reflection and/or transmittance by the recorded pits.

The optical disk and disk drive come into contact with the optical pickup. Figure 2 shows the relationship between the optical disk and the optical pickup. Light emitted from the semiconductor laser is collimated by the lenses, condensed by the objective lens, and then focused on the recording layer.

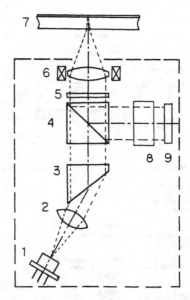

Figure 2. Relationship between optical memory disk and optical pickup. 1, Laser diode; 2, collimator; 3, wedge-mirror prism; 4, polarized beam splitter; 5, $\frac{1}{4}\lambda$ plate; 6, objective lens (focus actuator); 7, optical memory disk; 8, FO, TE detector; 9, photo (RF) detector.

Accordingly, the optical memory disk must feed the reflected light back to the optical pickup, where the reflected light must fulfill the focus controlling function as a focus error signal. Furthermore, the optical memory drive system must receive an address signal (PA signal) to correctly identify the positions of the optical memory disk and the track-crossing signal (TC signal) used to identify the direction of the movement of the optical pickup while the address searching operation is under way, as well as the tracking error signal (TE signal) used for controlling the optical pickup to correctly trace the predetermined tracks.

To achieve these objectives, the surface of the optical disk requires fine patterns to convert the radiated light on the disk into intensely or weakly reflected light which can then be properly modulated. A stamper is needed to produce these fine patterns.

A variety of fine patterns are classified into two types according to the method of generating the TE signal: the continuous servo (CS) system and the sampled servo (SS) system. One type of CS system provides continuous grooves (PG) on lines extended from PA pits. This is the system for recording information in the PG. In another type of CS system, PA pits are arranged between PG (called "land") so that recording pits can be located on the land.

Referring now to the system for recording information in the PG of the CS system, the method of externally delivering signals is described below.

Figure 3 is a diagram of the DC level electrical signal output and fine patterns. Using the CS system or the SS system, the intensity of light reflected from the flat area (called mirror portion) is maximized. The mirror portion

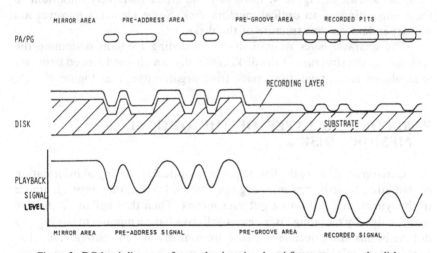

Figure 3. DC level diagram of reproduction signal and fine patterns on the disk.

is wider than the reading beam spot size (1–2 μm in diameter). The magnitude of the reflected light increases or decreases according to the disposition of the PA pits. Reflected light is modulated in this way so that address signals can be generated.

No modulation of reflected light takes place in the PG. The reflected light level of the PG is lower than that of the mirror portion. When information is recorded on the thin film in the PG, the recording pits function to decrease the amount of reflected light, thus generating reflection of light lower than that of the PG. In this way, addresses are generated between the mirror portion level and the PG portion level. Moreover, recorded signals are also generated lower than the PG level.

The land-recording format of the CS system maximizes the signal output from the independently disposed PA and PG portions. In the CS system, the difference between the magnitude of the reflected light from the PG and the magnitude of the reflected light from the land (which is substantially the track-crossing signal) plays an important role in controlling the disk-drive system.

The disk of the SS system does not have PG on the substrate. Instead, specially arranged pits called "wobble pits" play an important role in generating TE signals.

IV. STRUCTURE OF THE OPTICAL MEMORY DISK

Generally, conventional optical memory disks are composed of the elements shown in Figure 4. Each of these elements is very important to the composition of an optical memory disk. Some elements however are not dependent on the structure of the disks.

The characteristics desired of the recording medium determine the selection of the structure of the disk. Generally, an air-sandwiched structure is employed for optical disks made from organic dyes (see Figure 4).

V. RECORDING LAYER OF THE OPTICAL MEMORY DISK

Carlson *et al.*[9] were the first to propose optical recording of information by radiating laser beams onto organic material. First, they impregnated a triphenylmethane dye into a polymer matrix. Then they collapsed the dye in the matrix by radiating laser beams. All dyes have a number of conjugated double bonds and selectively absorb from ultraviolet to near-IR light. The organic dye with absorption matching the wavelength of the laser radiation

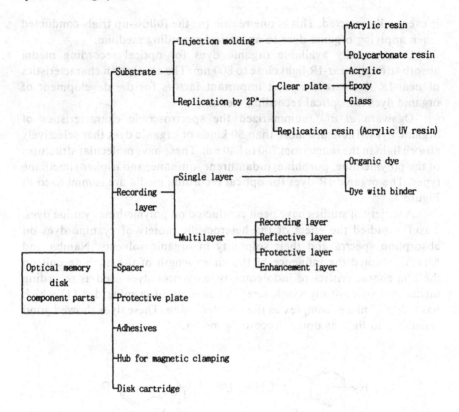

* 2P:Photopolymerization method

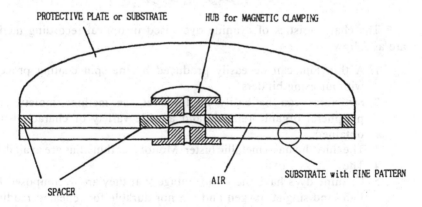

Figure 4. Components of the optical memory disk and typical disk structure.

is eventually selected. This is one reason for the follow-up trials conducted when applying organic dyes to an optical recording medium.

Commercially available organic dyes for optical recording media absorb strongly near-IR light close to 800 nm. The absorption characteristics of near-IR light are the most important factors for the development of organic dyes for optical recording media.

Ogawara *et al.*[10] summarized the spectroscopic characteristics of organic dyes. There are more than 50 kinds of organic dyes that selectively absorb light in the range from 700 to 840 nm. These have molecular structures of the polymethine, porphine, indanthrene, quinone, and triphenylmethane types. The organic IR dyes for optical recording media are summarized in Figure 5.

A variety of studies have been conducted on polymethine cyanine dyes. Yasui[11] studied the effect of the heterocyclic moiety of cyanine dyes on absorption spectra and their solubility in organic solvents. Namba and Matsui[8] studied the influence of the chain length of the methine unit on the film characteristics of indolenine type cyanine dyes used as recording media and successfully synthesized a new type of cyanine dye (1) which has a dithiol nickel complex as the counter anion. These dyes showed good resistance to light as optical recording media.

The characteristics of cyanine dyes used in optical recording media are as follows.

1. A thin film can be easily produced by the spin coating process without using binders.
2. Structures can be easily modified at the methine moiety and peripheral organic radicals and a wide variety of characteristics obtained.
3. The thin film has a metallic luster. Monolayer thin films are available.
4. They are nontoxic.
5. Cyanine dyes have the disadvantage that they are decomposed by light and singlet oxygen and are not durable for repeated reading use, as described in Section VII.D.

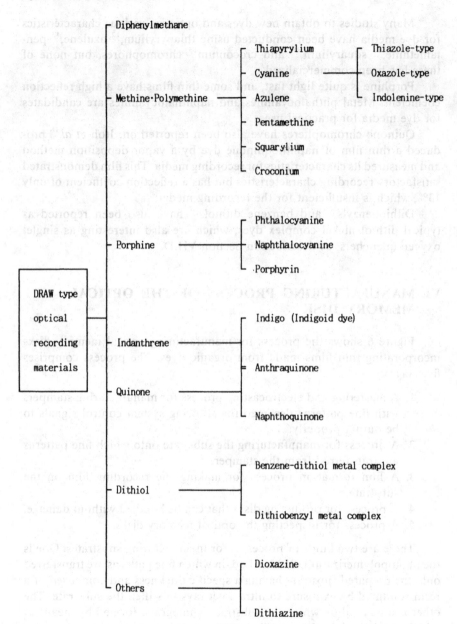

Figure 5. Organic dye materials for the optical recording layer.

Many studies to obtain new dyes and optimize various characteristics for dye media have been conducted using thiapyrylium,[12] azulene,[13] pentamethine,[14] squarylium,[14] and croconium[15] chromophores, but none of these has been commercialized.

Porphine is quite light fast, and some thin films have a high reflection coefficient. Metal phthalocyanines and naphthalocyanines are candidates for dye media for practical use.

Quinone chromopheres have also been reported on. Itoh et al.[16] produced a thin film of naphthoquinone dye by a vapor deposition method and measured its characteristics for recording media. This film demonstrated satisfactory recording characteristics but has a reflection coefficient of only 13%, which is insufficient for the recording media.

Dithiobenzyls[17] and benzene dithiols[18] have also been reported as typical dithiol nickel complex dyes, which are also interesting as singlet oxygen quenchers, as described in Section VII.D.

VI. MANUFACTURING PROCESS OF THE OPTICAL MEMORY DISK

Figure 6 shows the process for manufacturing optical memory disks incorporating thin films made from organic dyes. The process comprises five stages:

1. A mastering and electrocasting process for manufacturing stampers with fine patterns designed for allowing system control signals to be output properly.
2. A process for manufacturing the substrate onto which fine patterns are transferred from the stamper.
3. A film formation process for making the recording film on the substrate.
4. A process for producing disks that can be handled without damage.
5. A process for inspecting the optical memory disks.

There are two kinds of processes for manufacturing substrates. One is the photopolymerization (2P) method, in which fine patterns are transferred onto the prepared substrate having a specific thickness and composed of a resin solidified by exposure to ultraviolet rays to stiffen the substrate. The other is a method by which the substrate is integrally formed by means of a high-precision injection molding process.

Conventionally, the spin coating process is used for producing the thin film on the substrate. First, the organic dye for the optical recording medium is dissolved into a selected solvent at a predetermined concentration. The

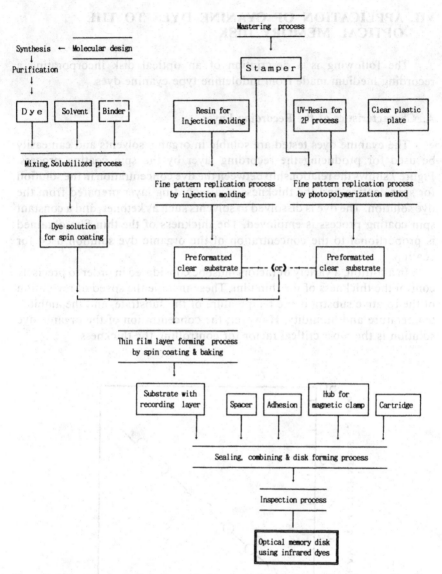

Figure 6. Process for manufacturing optical memory disks incorporating thin recording film made from organic dye.

prepared solution is then dropped onto the substrate, which is rotated at extremely high speed. The surplus solution is removed from the rotating substrate to complete the spin coating process. The disk manufacturing operation must be executed in a clean room.

VII. APPLICATION OF CYANINE DYES TO THE OPTICAL MEMORY DISK

The following is a description of an optical disk incorporating a recording medium made from indolenine type cyanine dyes.

A. Characteristics of the Recording Layer

The cyanine dyes tested are soluble in organic solvents and can easily be used for producing the recording layer by the spin coating process. Figure 7 shows the relationship between the dye concentration in the solution for spin coating and the thickness of a recording layer prepared from the dye solution. The dye is dissolved in solvents such as ketones, and a constant spin coating process is employed. The thickness of the thin film obtained is proportional to the concentration of the organic dye solution used for coating.

In addition, a variety of factors must be considered in order to precisely control the thickness of the thin film. These include the speed of revolution of the rotative substrate, the temperature of the substrate, and the ambient temperature and humidity. However, the concentration of the organic dye solution is the most critical factor for controlling the thickness.

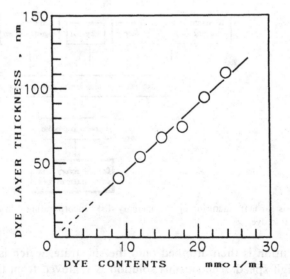

Figure 7. Relationship between concentration of the organic dye solution available for spin coating and thickness of thin film.

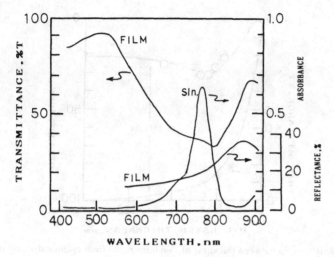

Figure 8. Spectroscopic characteristics of indolenine-type cyanine.

Figure 8 shows the spectroscopic characteristics of the cyanine dye used. The absorption maximum of the film exhibited a 40-nm bathochromic shift compared with that of the solution, and the film absorbs at 800 nm, which is the most favorable wavelength for laser diodes. Although the absorption maximum is quite important for sensitivity, the reflection coefficient is also important from the viewpoint of output of control signals. Figure 8 also represents the dependence of the reflection coefficient of the recording film on the wavelength.

The relationship between the reflectance (R), transmittance (T), and absorption coefficient (A) of the dye film at a given wavelength is expressed by Eq. (1).[19]

$$R + T + A = 1 \qquad (1)$$

A design which lowers the transmittance and increases the reflection coefficient and absorption is desirable. Needless to say, in addition to other characteristics inherent to the material, the thickness of the dye film is also very important. Optimum thickness allows the material to fully generate its best characteristics. Figure 9 indicates the relationships between the thickness of thin films made from cyanine dye and the reflection coefficient, transmittance, and absorption. Circles on the reflectance curve denote the experimental values. Transmittance decreases with increasing film thickness. The reflection coefficient is maximum at a cyanine dye film thickness of 700 Å and gradually decreases beyond this value of the film thickness.

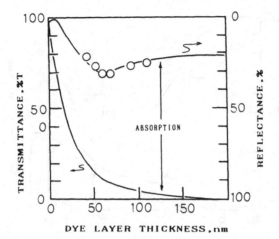

Figure 9. Relationship between thickness of thin films made from cyanine dye and reflection, transmittance, and absorption. Substrate (PMMA): $n = 1.48$, $k = 0.0$; dye layer (cyanine): $n = 2.70$, $k = 1.70$; air: $n = 1.00$, $k = 0.0$; $\lambda = 830\,nm$.

B. System Control Signal

1. Output of Reproduction Signal from Pre-pits

To control the optical memory disk system, all conventional optical memory disks contain a preformat part with address pre-pits (PP). These pits are present for all modulation methods and tracking control systems.

The magnitude of the output signal or the level and quality of the reproduced signal is determined by factors such as width, length, depth, and edge shape of the PP and thickness of the recording thin film made from the cyanine dye. When applying a rectangular model for example, the level of the reproduced signal reaches a maximum if the pits have a depth which is one-quarter the wavelength of the laser diode light.[20] If the wavelength is 830 nm, 1400 Å of depth is obtained for substrates made from the acrylic resin, whereas 1300 Å of depth is obtained for substrates made from polycarbonate resin. It appears that the maximum level of the reproduced signal is obtained at a depth slightly deeper than the above. In other words, the wider the pit, the greater is the signal output.

Figure 10 shows that the width of the pit barely affects the reproduced signal output from PP. The numbers on the figure represent the depth of the PP. The length of the pits was selected to be 3 μm to eliminate the adverse effect of the Modulation Transfer Function (MTF). The thickness of the thin film made from the dye is approximately 700 Å.

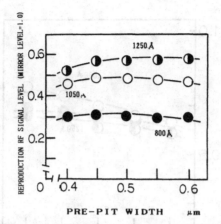

Figure 10. Influence of width of pits on reproduced signal output from PP.

If the shape of the pits is kept constant, then the thicker the thin film made from the organic dye, the greater is the signal output, and the signal waveform appears very strong and clear.

2. Tracking Error Signal Output

The output level of the tracking error signal depends on factors including width, depth of the PG, shape of the cross section of the PG, and thickness of the thin film. Figure 11 indicates an example of the effect of the width of the PG on the tracking error signal output level when the thickness of the thin film made from the dye is constant at 700 Å. The figure indicates that, at any depth of the PG, the output level of the TE signal increases with the width of the PG. The output level of the TE signal reaches a maximum when the width of the PG is one-half the track pitch.

Although, theoretically, the output level of the TE signal should reach a maximum when the depth of the PG is one-eighth the wavelength of the laser diode light (about 700 Å), in fact, it reaches a maximum at a point deeper than the theoretical value of the depth of the PG. This is probably because of the influence of the shape of the cross section of the PG, which is not perfectly rectangular.

Figure 12 shows the dependence of the output level of the TE signal on the thickness of the thin film made from the dye when the PG configuration remains constant. Given the thickness of the thin film and the reflection coefficient of the thin film as noted above, the output level of the TE signal appears to reach a maximum at the film thickness corresponding to the maximum value of the reflection coefficient.

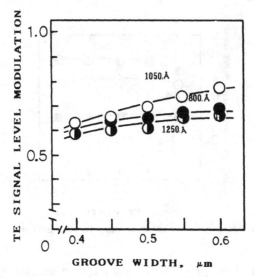

Figure 11. Influence of width of PG on TE signal output level when thickness of dye film is constant at 700 Å.

3. Track-Crossing Signal Output

Figure 13 shows the relationship between the TC signal level and the width of the PG. The spin coating conditions are constant, and the depth of that PG is shown in Figure 13 using 1.6 μm of track pitch. It is observed that the deeper the depth of PG, the higher the level of the TC signal. This

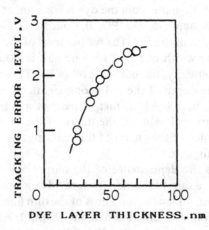

Figure 12. Dependence of output level of TE signal on thickness of thin film made from cyanine dye when the configuration of PG remains constant.

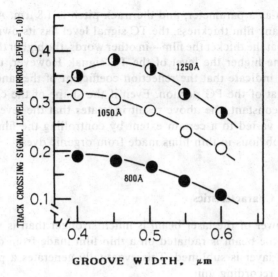

Figure 13. Relationship between TC signal level and width of PG when spin coating is under way under constant conditions.

is because there is a sizable difference between the dye film thickness of the land portion and that of the groove portion, resulting in a large difference in the magnitude of reflected light.

Figure 14 represents the level of the TC signal when the depth of the PG is constant and the width of the PG is varied. The thickness of the dye

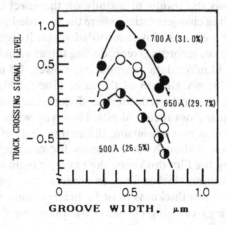

Figure 14. Level of TC signal when depth of PG is kept constant and width of PG is varied.

film functions as a parameter, and the track pitch is 1.8 μm. As shown in
the figure, at any film thickness, the TC signal level has its own peak. It is
understood that the thicker the film—in other words, the higher the reflection
coefficient—the higher the level of the TC signal. However, the negative
values shown indicate that the reflection coefficient of the land portion is
lower than that of the PG portion. Even if the shape of the cross section
of the PG is constant, the above result indicates that the level of the TC
signal can be varied to a certain extent by controlling the film thickness.
This is quite obvious in thin films made from organic dyes.

C. Recording Characteristics

If the power of the laser beam is much stronger than is required for
reading, and the beam is radiated on a thin film made from organic dye,
the recording layer is sublimed. This eventually generates a pit, which is
the minimum recording unit.

Many factors affect the recording characteristics. Those related to the
disk drive include recording power, density of beam energy, linear velocity,
and the shortest beam radiating time determined by the modulation format
(i.e., maximum frequency and shortest pit length). On the other hand, the
factor related to the disk itself is the thickness of the dye film at the recording
portion.

1. Effect of Thickness of Dye Film on Recording Characteristics

Figure 15 shows the results of a study on the effect of thicknesses of
dye films on recording characteristics, where the recorded pit length is 1 μm.
The test conditions were as follows: recorded signal frequency, 5.5 MHz at
1 : 1 of duty cycle ratio; recording/reproducing linear velocity, 11 m/s; peak
recording power, 11 mW; data reproducing power, 0.5 mW; NA of the
objective lens, 0.523; wavelength of the laser diode, 830 nm. The thicker
the film, that is, the higher the reflection coefficient, the greater is the output
level of the RF signal from recorded pits. However, when the thickness of
the recording thin film exceeds 60 nm, the output level of the reproduced
RF signal decreases. Although the efficiency for catching energy may be
raised by increasing the film thickness, the energy caught by the film is not
sufficient to fully sublime thicker films. In other words, optimum recording
power matching the film thickness must be present. This, in turn, suggests
that raising the energy-catching efficiency by increasing the film thickness
does not always work to increase the sensitivity.

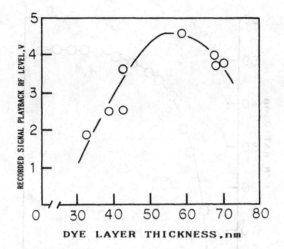

Figure 15. Example of result of study on influence of thickness of film made from cyanine dye on recording characteristics.

2. Dependence of Recording Signal on Frequency

Figure 16 shows the dependence of the carrier-to-noise (C/N) ratio on frequency. Test conditions were as follows: single-frequency recording signal; recording/reading linear velocity, 11 m/s; light wavelength, 830 nm; NA of the objective lens, 0.523. The number on the figure indicate peak

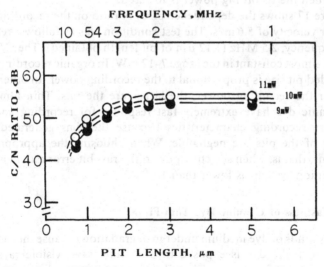

Figure 16. Dependence of C/N ratio on pit length.

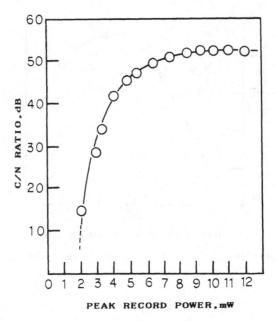

Figure 17. Dependence of C/N ratio on recording power at 5.6 m/s of linear velocity.

recording power. About 50 dB of C/N ratio was produced by applying approximately 1 μm of pit length. However, the C/N ratio decreases by 2 to 3 dB when the recording power is lowered.

Figure 17 shows the dependence of C/N ratio on the recording power at a linear velocity of 5.6 m/s. The test condition was as follows: recording signal frequency, 2.5 MHz (1.12 μm of pit length obtained). The C/N ratio remained almost constant in the range 7–12 mW. In organic recording media, the recorded pit size is proportional to the recording power. In other words, the higher the recording power, the larger are the pits. Thin films made from organic dye have extremely fast response for recording and other satisfactory recording characteristics because the rims generated in the periphery of the pits are negligible. When choosing the appropriate pit dimensions, that is, when selecting power, the raw bit error rate in a variety of modulation systems is lower than 10^{-6}.

D. Lightfastness of Cyanine Dye Thin Films

Many kinds of dye medium undergo degradation because the molecules collapse when the dye is exposed to ultraviolet rays, visible rays, near-infrared rays, or infrared rays. Thin films for optical recording media made from organic dyes are constantly exposed to near-infrared rays used for recording and reading signals. After repeated and intense exposure to the high-density energy of near-infrared rays focused by lenses, thin films undergo severe photodegradation and, as a result, lose their capability to function as recording films. Specifically, if a track is read continuously over a period of time, the written data will be destroyed or the PP will no longer be able to generate control signals. The durability of a track toward repeated reading operations during the data reproduction process involves a complex relationship between the magnitude of light and heat generated for accessing (i.e., the intensity of light for reading), density of optical energy, linear velocity for reading, thermal conductivity of the substrate and the thin film, light absorption, and so on. Figure 18 shows an example of these effects.

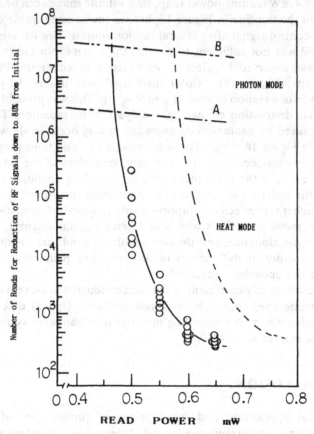

Figure 18. Relationship between intensity of light for reading data and rounds of repeated data reading operations needed for level of reproduced RF signal output from recorded signal to be reduced to 80% of the initial value.

The horizontal axis denotes the intensity of light for reading and vertical axis denotes the rounds of reading operations at which the RF signal output reaches 80% of the initial value, where the data are not yet destroyed. When the intensity of the reading light is lowered, the rounds of reading greatly increase. Some morphological changes in the recorded pits were observed due to the exposure to intense reading light, and "heat-mode degradation" was proposed.

When the data reproduction linear velocity is increased at constant light intensity, the rounds of reading increase. The dashed line in Figure 18 shifts to the right. Consequently, heat-mode degradation can be suppressed by selecting the light intensity which maintains the lowest speed of reading linear velocity.

When 0.4 mW reading power is applied, infinite rounds can be achieved as shown on the test line in Figure 18, but we found undesirable effects in the system control signal after several million rounds. The RF signal level from the PP was not sufficient to control the system. No morphological variation was observed by electron microscopy in either the PP or the recording pits (Figure 19). "Color discharge" was observed by optical microscopy in one section of about 1 μm in width. This was probably caused by molecular destruction of the organic dye film by radiation. The color discharge caused by photon-mode degradation may occur as shown by the chain line in Figure 18. Degradation of organic dye thin films prepared by CS and SS systems occurred competitively, regardless of the intensity of light. Differences in the principles of the data reading operation probably determine the order by which one system precedes the other.

A quencher very effectively suppresses photon-mode degradation. Conventionally, metal complexes with acetylacetonate, dithiobenzyl, benzene dithiol, salicylic aldoxime, and thiobisphenolate ligands are the quenchers available. In order to shift line A to or above line B in Figure 18, it is essential to use quenchers effectively.

A wide variety of experiments have been conducted to prevent degradation of cyanine dyes caused by exposure to light.[8] Optical disk systems which effectively combine recording material with hardware systems have been on the market since 1985.

VIII. CONCLUSION

Practical application and development of further uses of optical memory systems are currently under way. International standardization of DRAW type optical memory disks has been promoted since 1985. Standardization of data-erasable types of optical disks is also under consideration.

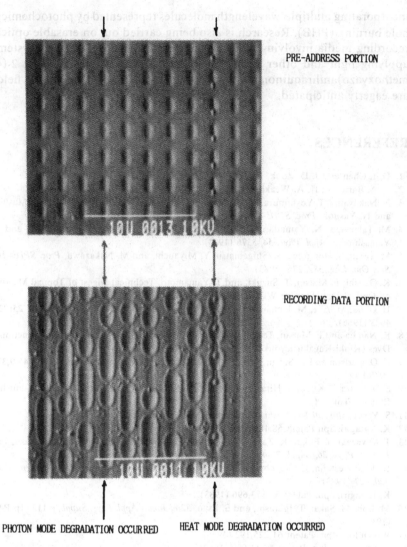

PRE-ADDRESS PORTION

RECORDING DATA PORTION

PHOTON MODE DEGRADATION OCCURRED   HEAT MODE DEGRADATION OCCURRED

Figure 19. SEM of photon-mode degradation and heat-mode degradation parts.

Efforts by all concerned are steadily bearing fruit for a stable and quantitative supply of optical disks at low cost. Furthermore, the ISO standardization of optical disks 130 mm in diameter is already under way[21] so as to render them completely interchangeable between systems.

Much research on organic optical recording media to obtain better functionality is in progress. This includes, for example, work on memory incorporating multiple-wavelength molecules represented by photochemical hole burning (PHB). Research is also being carried out on erasable optical recording media involving photochromism and on novel memory systems applying light and other energy in harmonious combination using 2-(4-methoxyazo)anthraquinone.[22] Further developments in these novel fields are eagerly anticipated.

REFERENCES

1. D. I. Chen and J. D. Zook, *Proc. IEEE 63*(8), 1207 (1975).
2. R. A. Bartolini, H. A. Weakliem, and B. F. Williams, *Opt. Eng. 15*(2), 99 (1976).
3. N. Nakayama, T. Yoshimura, and C. Nagou, *Toshiba Review 36*(3), 250 (1981): M. Mashita and N. Yasuda, *Proc. SPIE-Int. Soc. Opt. Eng. 329*, 190 (1982).
4. M. Takenaga, N. Yamada, K. Nishiuchi, N. Akahira, T. Ohta, S. Nakamura, and T. Yamashita, *J. Appl. Phys. 54*, 5376 (1983).
5. M. Terao, S. Horigome, K. Shigematsu, Y. Miyauchi, and M. Nakazawa, *Proc. SPIE-Int. Soc. Opt. Eng. 382*, 276 (1983).
6. K. Ogoshi, F. Matsui, T. Suzuki, and T. Yamamoto, Technical Digest of Topical Meeting on Optical Data Storage WDD2, Washington, D.C. (1985).
7. H. Ohba, M. Abe, M. Umehara, T. Satoh, Y. Ueda, and M. Kunikane, *Appl. Opt. 25*(22), 4023 (1986).
8. K. Namba and F. Matsui, Technical Digest of the Symposium on Chemistry of Functional Dyes (Kinki Kagaku Kyoukai, Osaka, Japan), No. 6 (1988).
9. C. O. Carlson and E. Storn, *Science 154*, 1550 (1966); C. O. Carlson, Jpn. Patent 45 9,333 (1970).
10. S. Ogawara, T. Kitao, K. Hirashima, and M. Matsuoka, eds., *Organic Colorants*, Kodansha-Elsevier, Tokyo (1988).
11. S. Yasui, *Shikizai Kyokaishi 60*, 212 (1987) (in Japanese).
12. K. Katagiri, Jpn Patent 58 181,688 (1983).
13. T. Miyazaki, T. Fukui, K. Takano, Y. Oguchi, T. Santoh, K. Nishide, and Y. Takasu, *Jpn. J. Appl. Phys. 26, Suppl. 26-4*, 33 (1987).
14. D. J. Gravesteijn, C. Steenbergen, and J. Van Der Veen, *Proc. SPIE-Int. Soc. Opt. Eng. 420*, 327 (1983).
15. K. Katagiri, Jpn. Patent 58 173,696 (1983).
16. M. Itoh, M. Sakai, T. Igarashi, and S. Esyo, *32nd Japan Appl. Phys. Suppl.*, p 112 (1p-P-9) (1985).
17. W. Shulott, Jpn. Patent 61 225,192 (1986).
18. Y. Sasagawa, Jpn. Patent 57 11,090 (1982).
19. H. Ohba, Ricoh Technical Report, No. 13, p. 46 (1985).
20. H. H. Hopkins, *J. Opt. Soc. Am. 69*(1), 4 (1979).
21. F. Matsui, S. Yanagisawa, T. Miyadera, K. Nanba, and T. Aoi, *Shingakugihou 87* (387) 21 MR87-53 (1988) (in Japanese).
22. T. Shimizu, 2nd International Symposium on Bioelectronic and Molecular Electronic Devices, R&D Association for Future Electron Devices, pp. 151-154, Dec. 12-14, 1988, Fujiyoshida, Japan.

11
Thermal Writing Displays

YOSHIHARU NAGAE

I. INTRODUCTION

Thermally addressed displays, where the heat input is provided by a laser beam or a heat pulse current, have been established for several years.[1-21] It is generally known that the thermal writing liquid-crystal display is the most promising because of its intrinsic memory effect. The display overcomes the inherent addressing limitation associated with liquid-crystal displays that use just an electro-optical effect because that effect lacks memory and has insufficient nonlinearity to keep a large number of lines separate.

When a laser beam is used to write information onto a liquid-crystal panel, the optical energy of the laser beam should be effectively converted to thermal energy. In order to increase writing speed and to decrease laser power, a special thin absorbing layer made of some metal oxides or organic materials is designed. Infrared absorbing dye molecules are introduced for the same purpose.

By using the thermo-electro-optical effects of smectic liquid-crystal (LC) materials, two types of display systems have been developed, namely, a projection display and a flat panel display. These displays are described in this chapter.

YOSHIHARU NAGAE • Hitachi Research Laboratory, Hitachi Ltd., Hitachi, Ibaraki 319-12, Japan.

II. PROJECTION DISPLAY

A. Liquid-Crystal Light-Valve Cell

Figure 1 shows a schematic diagram of a reflective liquid-crystal light-valve cell.[20] Laser diodes are used for thermal addressing. Biphenyl smectic-A liquid-crystal material is sandwiched between two glass substrates. On the inner surfaces of the substrates, alignment layers are placed to achieve homeotropic structure. On the left-hand side of the LC layer, a Cr_2O_3 laser-beam-absorbing layer and Al reflector are placed between the glass substrate and the alignment layer. The Al reflector is used as an electrode as well. On the opposite side, a transparent electrode (ITO) is placed between the glass substrate and the alignment layer. The outer surfaces of the substrates are covered with antireflection (AR) layers for laser beam and projection light.

Thermal addressing on the liquid-crystal light-valve cell is performed as follows. For thermal addressing, the laser beam is focused on the absorbing layer from the left and scanned from top to bottom. Thermal energy is absorbed by the Cr_2O_3 layer, which heats the adjacent LC from the smectic-A phase into the isotropic phase. In the cooling period that follows this heating process, the LC structure changes to the focal-conic structure with scattering domains when no video signal voltage is applied. On the other hand, the

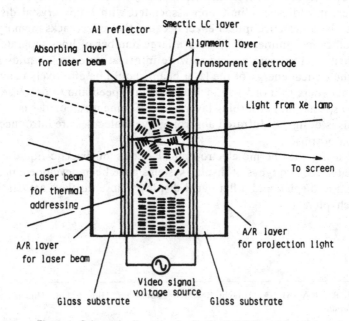

Figure 1. Schematic diagram of reflective LC light-valve cell.

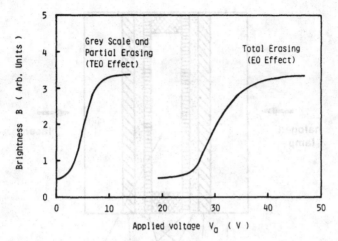

Figure 2. Brightness B of projected image on the screen as a function of applied voltage V_a. TEO, thermo-electro-optical; EO, electro-optical.

LC structure returns back to the initial homeotropic structure when a video signal voltage of about 10 V is applied. When the video signal voltage is between 0 and 10 V, a focal-conic structure, with varying size and density of the scattering domains, is formed. Thus, by controlling the video signal voltage, reflectance of the LC cell can be controlled, and therefore the intensity of the reflected light to be projected onto the screen can also be controlled. With this procedure, gray-scale images can be obtained on the screen.

Figure 2 shows brightness B of projected images on the screen as a function of applied video signal voltage V_a.[20] For thermal addressing, the thermo-electro-optical (TEO) effect of smectic-A LC material is employed to obtain gray-scale images and to partially erase. The B-V_a characteristics for this effect are shown on the left of the figure. On the other hand, for total erasing the electro-optical (EO) effect is used, and the B-V_a characteristics are shown on the right of the figure. Since the voltage ranges of these two effects are separated, thermal addressing and total erasing can be performed independently.

Urabe *et al.* reported another structure of a reflective type liquid-crystal light-valve cell, shown in Figure 3, in which an infrared absorbing dye is doped into smectic-A LC material.[14] The cold mirror and cold filter layers are dielectric multilayers coated on borosilicate glass substrates. The cold mirror layer has 92% transmittance at 780 nm and it totally reflective between 415 and 600 nm, so that a laser beam with 780-nm wavelength strikes the LC layer, where the laser energy is absorbed by the incorporated squarylium dye. The cold filter is highly reflective at 780 nm and has 90% transmittance

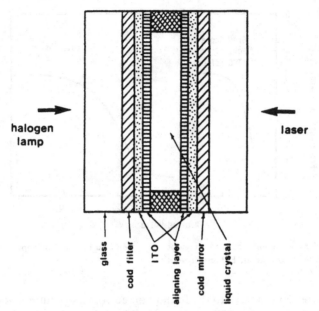

Figure 3. LC cell structure.

in the 415–660-nm visible range, so that the part of the laser beam which was not absorbed is reflected and strikes the LC layer again. A projection light passes through the cold filter layer and is reflected by the cold mirror layer.

Urabe *et al.* made the following comments concerning infrared dyes.[14] Among various types of infrared dyes, the most common are cyanine dyes, which are known as laser dyes for infrared lasing. Generally, they have high absorption coefficients and high dichroic ratios, which are desirable characteristics for this application, and some of them can be dissolved in the LC materials. However, their lack of durability is a serious problem. Furthermore, Urabe *et al.* observed that the alignment of LC material is destroyed by decomposed cyanine dye. Some metal chelate compounds have absorption in the infrared range and show very high solubility in LC material. However, the absorption coefficients of these type of dyes are small, so that large amounts of dye would have to be dissolved, which would affect the physical properties of the host LC material. Taking these facts into consideration, Urabe *et al.* chose a squarylium dye. The squarylium dye is more durable than cyanine dyes. Moreover, it has similar absorption characteristics to the cyanine dyes.

The absorption spectrum of squarylium dye NK-2772 (Nippon Kanko Shikiso Kenkyusho Co.) is shown in Figure 4. There is weak absorption in

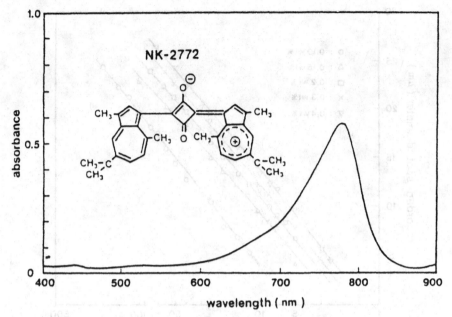

Figure 4. Absorption spectrum of 0.2 wt % NK-2772 in CNB in isotropic phase (cell thickness, 12 μm).

the visible range due to absorption by the azulene rings. The absorption maximum is 780 nm, which is close to the wavelength of the GaAlAs laser used in the projection display. The molar absorption coefficient of NK-2772 is 1.15×10^5 dm^3 mol^{-1} cm^{-1}, and the dichroic ratio in smectic LC is 6.4.

Figure 5 shows plots at various dye concentrations of recorded spot size versus laser pulse width, with a cell thickness of 9 μm. As is evident from this figure, high concentration of dye causes high recording speed, and thinner cells have a faster speed.

Sasaki *et al.* also reported a similar idea to increase the addressing speed of a smectic–cholesteric light valve used in a projection display.[16] In their case, they used dichroic dyes with absorption maxima in the visible wavelength range because they choose an Ar laser and a He–Ne laser as the addressing laser beam source.

B. Optical System of the Projection Display

The optical system of the eight-color projection display is schematically shown in Figure 6.[21] The system is composed of a laser scanning system, smectic-A phase liquid-crystal light-valve cells, and a projection system. To

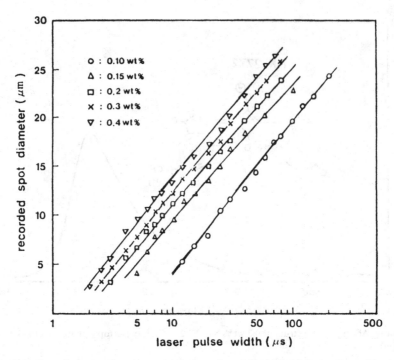

Figure 5. Recorded spot diameter versus laser pulse width (cell thickness, 9 μm; bias temperature, 4°C).

obtain eight-color images, three liquid-crystal light-valve cells corresponding to the three primary colors are installed. When three independent scanning systems for the three primary colors are used to address laser beams to the liquid-crystal cells, it is impossible to achieve good registration because the linearity of galvano mirror scanners is insufficient. Therefore, only one set of XY galvano mirror scanners is employed. In the laser scanning system, beams from two laser diodes are combined into one beam by a polarizing prism. The combined beam is scanned in the x, y directions by XY galvano mirror scanners. After passing through the f–θ lens, the laser beam is divided into three parts, each having the same power, by a beam splitting system. The divided laser beams are addressed to the three liquid-crystal cells.

The projection system is composed to one 2-kW Xe lamp, one projection lens, two dichroic prisms, and a screen. Light from the Xe lamp is divided by two dichroic prisms into three portions and projected onto the liquid-crystal cells. Reflected light from the liquid-crystal cells is combined by the same dichroic prisms as before and projected onto a 2 m × 2 m screen through the projection lens.

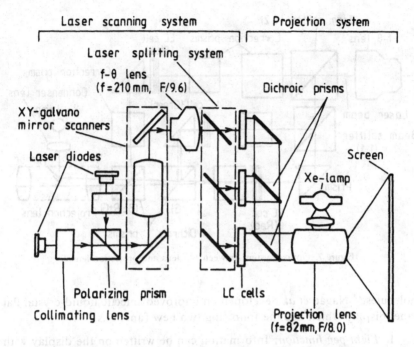

Figure 6. Optical system of prototype multicolor projection display.

The precise optical system between the $f-\theta$ lens and the projection lens is shown in Figure 7.

C. Displayed Images

Specifications of the right-color and full-color display described in this chapter are listed in Table 1.[20] Photographs of the displayed images are shown in Figures 8 and 9.

III. FLAT PANEL DISPLAY

A. Display Panel and Write-In Function

Hareng *et al.* succeeded in applying the same thermo-electro-optical effect of smectic-A LC material to a direct-view flat panel display by using heating electrodes.[12] Lu *et al.* have improved the viewing-angle characteristics of this type of display by mixing in a small amount of pleochroic-dye

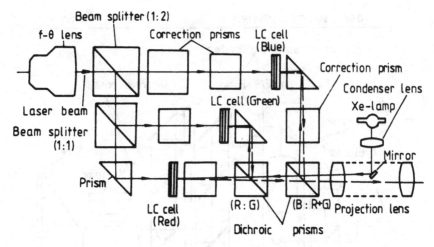

Figure 7. Optical system between f-θ lens and projection lens.

molecules.[13] Nagae *et al.* developed an improved smectic liquid-crystal flat panel display which has the following two new functions[15,17]:

1. *Light-pen function*: Information can be written on the display with a light pen.
2. *Read-out function*: The manually written information can be read out electrically.

Figure 10 shows a schematic diagram of a panel structure and of the addressing mechanism. Addressing can be performed by both heating pulses and a laser beam. In the usual smectic liquid-crystal flat display, the current

Table I. Specifications of Prototype Display

Screen size	2 m × 2 m
Number of	1000 × 1000
addressable pixels	(full color with raster scan)
	2000 × 2000
	(multicolor with vector scan)
LC cell size	20 mm × 20 mm
Laser power	12 mW at each LC cell
Writing speed	1 min/screen
	(full color with raster scan)
	30 m s^{-1}
	(multicolor with vector scan)
Brightness	30 ft-lambert (screen gain = 3.5)

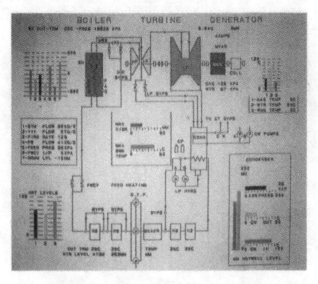

Figure 8. Photograph of displayed multicolor image.

Figure 9. Photograph of displayed full-color image.

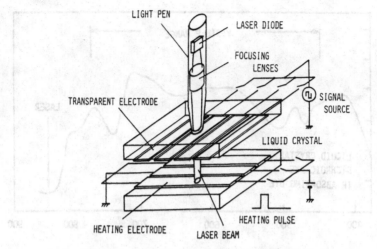

Figure 10. Schematic diagram of addressing mechanism.

pulses are applied to the heating electrodes so that the liquid-crystal material near the heating electrodes heats up to the clearing temperature. Signal voltage is then selectively applied to the pixels in the cooling period after cutting off the heating pulse. The pixels with applied signal voltage correspond to unwritten pixels, since they remain transparent. On the other hand, the pixels with no voltage corrspond to written pixels, since light is scattered strongly. With this method, any information can be displayed on the display in a line-at-a-time sequence.

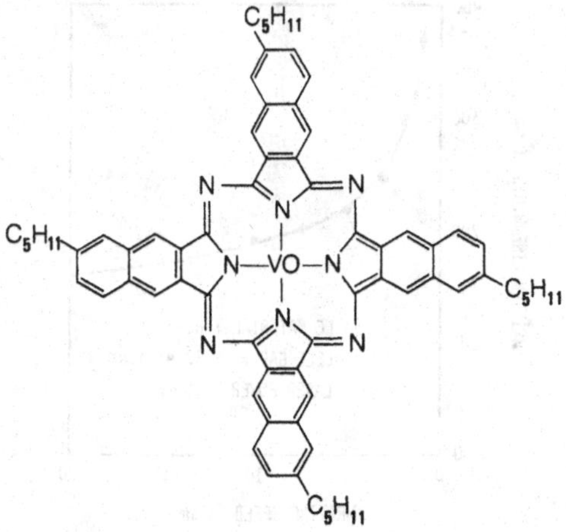

Figure 11. Molecular structure of naphthalocyanine dye.

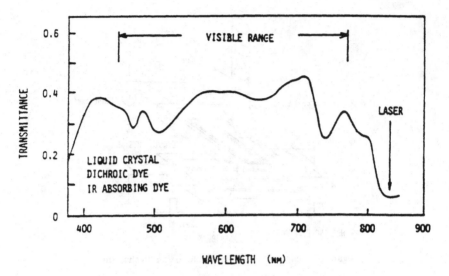

Figure 12. Transmission spectrum of LC panel.

Furthermore, in the developed display, a light pen is effectively used to manually input any information on the display panel. The light pen is equipped with a laser diode and focusing lenses. The light pen is moved to the place in the display panel where relevant information should be written. The emitted laser beam is focused on the heating electrodes, generating heat locally. The heat is transferred to the liquid-crystal layer

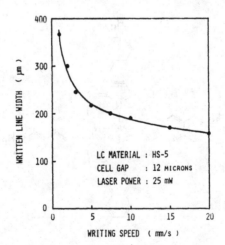

Figure 13. Writing characteristics.

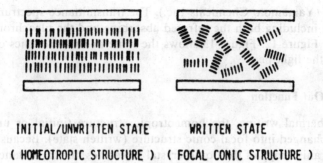

INITIAL/UNWRITTEN STATE   WRITTEN STATE
(HOMEOTROPIC STRUCTURE)   (FOCAL CONIC STRUCTURE)

Figure 14. Molecular alignments of pixel state.

near the heating electrodes, resulting in the change of the liquid crystal to the isotropic state. During the cooling period, the focal-conic structure, which has a strong scattering characteristic, is formed.

Nagae *et al.* also reported the use of an infrared absorbing dye to improve the writing speed of the light-pen function.[19] They also used a pleochroic dye to improve viewing-angle characteristics as was done by Lu *et al.* They chose the naphthalocyanine dye shown in Figure 11, named

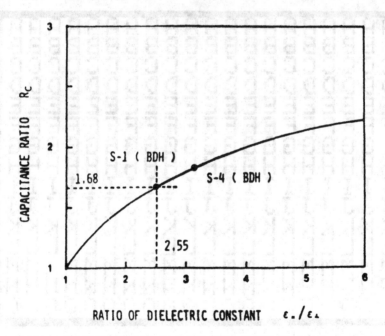

Figure 15. Dependence of capacitance ratio R_c on $\varepsilon_{\parallel}/\varepsilon_{\perp}$.

MB-IR-1 (Yamamoto Chemicals Inc.). The transmittance spectrum of the LC panel including both the infrared absorbing and the pleochroic dye is shown in Figure 12. Figure 13 shows the write-in characteristics obtained by using the light pen.

B. Read-Out Function

By thermal writing, the homeotropic structure (initial or unwritten state) is changed into focal-conic structure (written state). Because of the dielectric anisotropy of the liquid-crystal molecules, the capacitance of the written pixel is different from that of the unwritten pixel. As shown in Figure 14, it can be assumed that the molecular alignment of the unwritten pixel is perfectly perpendicular to the boundary and that of the written pixel is perfectly random.[17]

According to this assumption, the capacitance of the initial or unwritten pixel C_{uw} and that of the written pixel C_w are proportional to $\varepsilon_{\parallel}$ and $(\varepsilon_{\parallel} + 2\varepsilon_{\perp})/3$, respectively, where $\varepsilon_{\parallel}$ is the dielectric constant parallel to the long axis of the liquid-crystal molecules and $\varepsilon_{\perp}$ is the value perpendicular

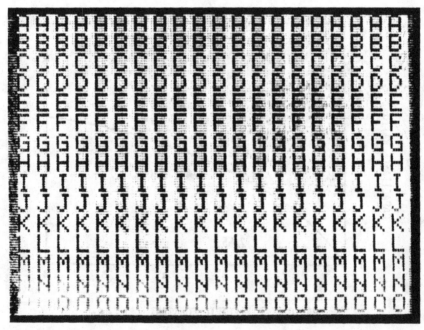

Figure 16. Photograph of displayed characters generated by electrode-heating method.

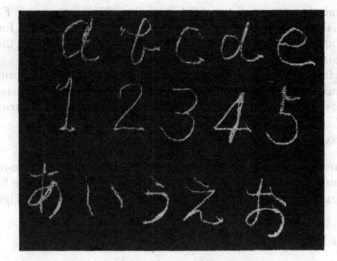

Figure 17. Photograph of displayed characters written by light pen.

to the long axis. The ratio of these capacitances, R_c, is given by

$$R_c = \frac{C_{uw}}{C_w} = \frac{3(\varepsilon_\parallel/\varepsilon_\perp)}{(\varepsilon_\parallel/\varepsilon_\perp) + 2}$$

For the ideal case, $\varepsilon_\parallel \gg \varepsilon_\perp$, and the capacitance ratio becomes 3. The dependence of R_c on $\varepsilon_\parallel/\varepsilon_\perp$ is illustrated in Figure 15, where data for two

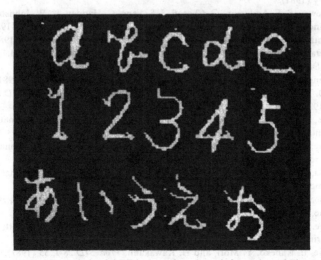

Figure 18. Photograph of CRT image of read-out information of Figure 17.

typical smectic materials S-1 and S-4 (both from BDH) are shown. For S-1, R_c is equal to 1.68 because $\varepsilon_\| = 12$ and $\varepsilon_\perp = 4.7$. It is found that for most smectic liquid-crystal materials, R_c is roughly equal to 1.7. Thus, the state of each pixel can be determined by measuring the capacitance.

Special circuit technology has been developed for measuring the capacitance of each pixel arranged in XY matrix form. By using this technology, the manually input information can be read out electrically.

C. Displayed Image and Read-Out Image

Figure 16 shows a displayed image written by the electrode-heating method, and Figure 17 shows the displayed image manually written by light pen.[17] Figure 18 shows the CRT image of read-out information of Figure 17.

REFERENCES

1. A. Sasaki, K. Kurahashi, and T. Takagi, Conf. Rec. IEEE Conf. Display Devices, p. 161 (1972).
2. D. Maydan, H. Melchior, and F. J. Kahn, Conf. Rec. IEEE Conf. Display Devices, p. 166 (1972).
3. M. Melchior, F. J. Kahn, D. Maydan, and D. B. Fraser, *Appl. Phys. Lett.* 21(8), 392 (1972).
4. D. Maydan, *Proc. IEEE 61*, 1007 (1973).
5. F. J. Kahn, *Appl. Phys. Lett.* 22(3), 111 (1973).
6. M. Hareng, S. Le Barre, and R. Hehlen, *Rev. Tech. Thomson-CSF 9*(2), 373 (1977).
7. A. Dewey, J. Jacobs, and B. Huth, *Proc. SID 19-1*, (1978).
8. R. Tsai, A High Data Density Multicolor Liquid Crystal Display, Technical Report, Singer Company (1981).
9. A. G. Dewey, S. F. Anderson, G. Cheroff, J. S. Feng, C. Handen, H. W. Johnson, J. Leff, R. T. Lynch, C. Marinelli, and R. W. Schmiedeskamp, *SID 83 Digest*, p 36 (1983).
10. K. Kubota, S. Sugama, S. Naemura, and N. Nishida, *SID 83 Digest*, p. 44 (1983).
11. K. Kubota, S. Naemura, S., Komatsubara, M. Imai, Y. Kato, N. Nishida, and M. Sakaguchi, *SID 85 Digest*, p. 260 (1985).
12. M. Hareng, S. Le Barre, R. Hehlen, and N. Perbet, *SID 81 Digest*, p. 106 (1981).
13. S. Lu, D. H. Davies, R. Albert, D. Chung, A. Hochbaum, and C. Chung, *SID 82 Digest*, p. 238 (1982).
14. T. Urabe, H. Usui, K. Arai, and A. Ohnishi, Proceedings of the 3rd International Research Conference (Japan Display), p. 486 (1983).
15. Y. Nagae, H. Kawakami, and E. Kaneki, Proceedings of the 3rd International Reserch Conference (Japan Display), p. 490 (1983).
16. A. Sasaki, N. Hayashi, and T. Ishibashi, Proceedings of the 3rd International Research Conference (Japan Display), p. 494 (1983).
17. Y. Nagae, H. Kawakami, and E. Kaneko, *IEEE Trans. Electron Dev. ED*-32, 744 (1985).
18. Y. Nagae, H. Kawakami, and E. Kaneko, *SID 85 Digest*, p. 289 (1985).
19. Y. Nagae, M. Takasaka, M. Kitajima, and H. Kawakami, in: Technical Report, Japan TV Engineering Association, ED892, p. 45 (1985).
20. Y. Nagae, E. Kaneko, Y. Mori, and H. Kawakami, *Proc. SID 28-1*, 55 (1987).
21. Y. Mori, Y. Nagae, E. Kaneko, H. Kawakami, T. Hashimoto, and H. Shiraishi, *Display*, 51 (April 1988).

12

Laser Printer Application

ATSUSHI KAKUTA

I. INTRODUCTION

Electrophotography has grown in the past two decades from office copier and duplicator applications to surpass other conventional printing methods because of its outstanding characteristics of high speed and high print quality. Now, laser printers and LED printers, based on the same printing principle of electrophotography, are widely used for computer printouts or multiple peripheral printers in office workstations and facsimiles.

The photoconductive material used as the photoreceptor plays an important role in the electrophotography printing system. In copiers or duplicators, the reflected light of the documents is used as a light source for the photoreceptor, and visible-light-sensitive materials, for example, amorphous selenium, zinc oxide, cadmium sulfide, or organic photoconductors, are adopted as the photoreceptors. However, in the case of laser printers, the light source is currently a GaAlAs diode laser which emits near-infrared light with a wavelength of around 800 nm, and special photoconductive materials sensitive to near-infrared wavelengths are needed. Organic materials have been considered as the most promising for this purpose since the beginning, because of their wide variety and the capability of introducing structural modifications to get desired properties. Many kinds

---

ATSUSHI KAKUTA • Hitachi Research Laboratory, Hitachi Ltd., Hitachi-shi, Ibaraki-ken 319-12, Japan.

of organic infrared absorbing compounds are presently applied as photoreceptors in laser printers, and some are introduced in this review.

II. ELECTROPHOTOGRAPHY

The principle of electrophotography and related topics are briefly introduced in this section. A more detailed treatment of electrophotography is available in ref. 1.

A typical electrophotographic system consists of five successive processes, as shown in Figure 1. At first, the photoreceptor, which is usually drum-shaped, is charged uniformly, at about 500–1000 V, on the surface in the dark. The photoreceptor is rotated at a constant speed, 10 to 50 rpm, while it undergoes the next process continuously. The charged photoreceptor is exposed to an imaging light, which is the reflection of the documents in the case of conventional copiers or the laser scanning light in printers and digital copiers. From this exposure, the energy of which is 5–30 erg cm^{-2}, an electrostatic latent image is formed on the photoreceptor surface and then visualized with developers of carbon or other tiny coloring particles in the next step. The visible image is transferred to a blank paper by another charging unit, and this transferred image is fixed by fusing to obtain a print. The photoreceptor used is ready for the next printing cycle without delay after a cleaning and discharging.

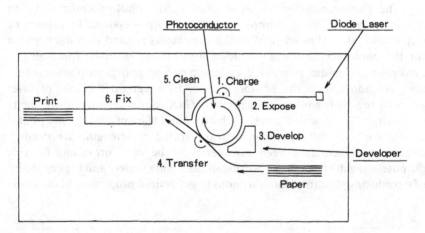

Figure 1. Laser beam printer. The photoconductive drum rotates while repeating processes 1 through 5; then prints can be obtained successively.

III. DUAL-LAYERED PHOTORECEPTOR

The first commercial organic photoconductor (OPC) was the poly(vinylcarbazole) and trinitrofluorenone complex material developed by IBM in 1969.[2] That was a sheet-shaped, mono-layered photoreceptor. Presently, dual-layered type OPCs are more popular since they have expanded capabilities.

As shown in Figure 2, the dual-layered photoreceptor is composed of a charge transport layer (CTL) and a charge generation layer (CGL), which are coated on a conductive substrate, one after another. The charging polarity is negative (minus) for this type of OPC to make better use of the high hole mobility of the CTL, as mentioned below. The imaging light exposure passes through the transparent CTL and reaches the CGL to generate charged photocarriers. Under a high external electric field, around 10^5 V cm^{-1} as applied by the first charging step, the charged pairs are effectively separated according to their polarities, and positively charged ones, that is, holes, are injected from the CGL to the CTL. The injected holes are moved by the external field to the surface to diminish the surface negative charge in the light-exposed area. The formation of an electrostatic latent image is thus accomplished. The photosensitivity of a photoreceptor can be measured in terms of the light energy required to decrease the surface voltage to a certain amount. Usually, the energy to decrease the surface voltage by half, E_{50} or $E_{1/2}$, is used as a standard parameter for photosensitivity. By adopting a dual-layered structure, two fundamental necessities for photoreceptors, charge generation and charge sustentation to form a

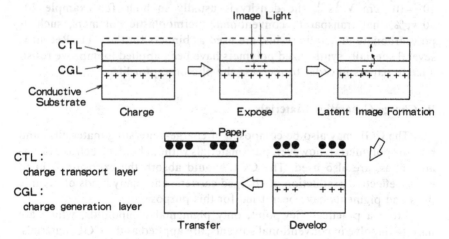

Figure 2. Electrophotographic printing process for the dual-layered photoreceptor.

sufficient electrostatic contrast, are shared by two different materials. The sensitivity of the photoreceptor mainly depends on three characteristics, namely, the charge generation efficiency of the CGL material, the charge transport efficiency or the carrier mobility of the CTL material, and the injection efficiency between the CTL and the CGL.[3,4]

A. Charge Transport Materials

The CTL is composed of charge-transporting material(s) and binding polymer(s), the thickness of which is 10–20 μm, while the CGL is usually less than 1 μm thick. Because of the fragility of the CGL, the CTL is usually used as the upper layer of the dual-layered structure, and it must be transparent to the exposing light wavelength. Although organic materials show both electron and hole mobilities, most electron-transporting materials, that is, electron acceptors, are not easily prepared as a uniform solid solution of polymers and, hence, are rarely applied as commercial photoreceptors.[5,6] In almost all CTLs, hole-transporting materials, that is, electron donors, are being used. Some of them are shown in Figure 3. Hydrazone derivatives or triphenylamines are popular in commercially available products. Each molecule in the figure has a long conjugated system in its structure along with electron-donating substituent(s). A significant point is these molecules have absorption edges at rather short wavelengths in spite of low ionization potential values. The hole mobility depends on the distance between the charge transport molecules in the layer, that is, the mixture ratio or the density of the charge-transporting materials in the binder.[7,10] To obtain a practically sufficient value for carrier mobility, 10^{-8}–10^{-4} cm^2 V^{-1} s^{-1}, the density is usually so high, for example 20–50 wt %, that transparent conventional thermoplastic polymers, such as polycarbonate or acrylic resins, are used as binders for the CTL. Recently, several specially synthesized polymers have been applied to improve resistance to papers in order to increase the photoreceptor life.

B. Charge Generation Materials

The CGL may also be composed of charge-generating material(s) and binding polymer(s); however, pure materials without binder, such as evaporated films, are also used. The CGL should absorb the exposure light to ensure effective production of charged carriers, and many kinds of organic dyes and pigments have been tried for this purpose.

From a practical viewpoint, only pigmental compounds, which are hard to dissolve in conventional solvents, are applied as the CGL materials. For the CGL, not only high quantum efficiency, or charge generation

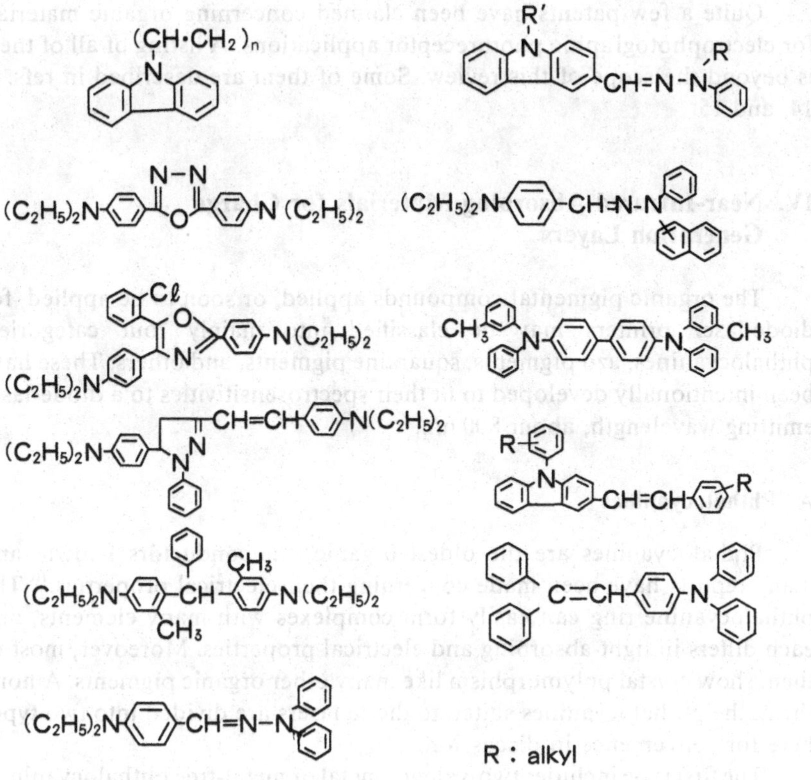

Figure 3. Some examples of charge-transporting materials. Typical materials proposed in patent applications are shown. An enormous number of other materials have been reported, and the list is actively being added to.

capability, is needed, but also prevention of unnecessary carrier injection from the conductive substrate, or carrier barrier formation to the substrate material, is necessary,[11] along with sufficient mobility of generated carriers. Since conventional organic pigments cannot easily achieve all of these CGL requirements, practically available pigments are limited. The CGL thickness is also restricted—typically it must be below 1 μm—for the same reasons. Polyazo pigments, perylene pigments, and others have been adopted as CGLs for copiers due to their strong absorptions in the visible region. Then the CTL should be selected afterward to provide a suitable energy level matching the CGL. This is critical for carrier injection efficiency from the CGL to the CTL.[11-13] This efficiency frequently dominates the entire sensitivity of the photoreceptor.

Quite a few patents have been claimed concerning organic materials for electrophotographic photoreceptor applications. A listing of all of them is beyond the scope of this review. Some of them are described in refs. 1, 14, and 15.

IV. Near-Infrared Absorbing Materials for Charge Generation Layers

The organic pigmental compounds applied, or soon to be applied, for diode-laser printers may be classified into mainly four categories: phthalocyanines, azo pigments, squaraine pigments, and others. These have been intentionally developed to fit their spectrosensitivities to a diode-laser emitting wavelength, about 800 nm.

A. Phthalocyanines

Phthalocyanines are the oldest organic semiconductors known, and many reports have been made concerning their electrical properties.[16] The phthalocyanine ring can easily form complexes with many elements, and each differs in light-absorbing and electrical properties. Moreover, most of them show crystal polymorphism like many other organic pigments. Among them, the phthalocyanines suited to diode lasers are divided into two types here for convenience in discussion.

The first type includes two-valence metal or metal-free phthalocyanines. Phthalocyanines of this type are used as conventional cyanic pigments, but special crystal forms that show near-infrared photoconductivity can be obtained. χ-form metal-free and copper phthalocyanines,[17,18] ε-form copper phthalocyanine,[19] and τ-form metal-free phthalocyanine[20] are reported to show characteristic strong absorption maxima in the near-infrared region and good photoconductivity attributed to the absorption. Aggregated magnesium[21,23] or zinc phthalocyanines[22,23] are included among them. All of these phthalocyanines show ordinary polymorphism; they crystallize in a thermodynamically unstable α-form and a stable β-form, both monoclinic. Each form is characterized by its visible absorption spectrum and X-ray diffraction pattern. The special forms are synthesized by a morphological transformation induced with organic solvents, for example, methylene chloride, and/or mechanical milling of α-forms. Excess treatment results in β-forms.

An example of a spectral photosensitivity plot is shown in Figure 4, where the dual-layered photoreceptor is composed of the metastable form, τ-form metal-free phthalocyanine, and a charge-transporting material, an

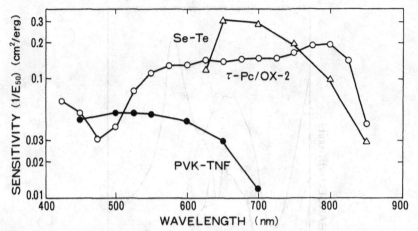

Figure 4. The spectral sensitivity of the dual-layered photoreceptor composed of τ-form phthalocyanine and a charge-transporting material.

oxazole derivative.[24] The sensitivity E_{50} is around 5 erg cm^{-2} at 800 nm (Figure 5), and the profile of the spectral dependence corresponds to the absorption spectrum of the phthalocyanine, which is shown in Figure 6. A monolayered photoreceptor in which ε-form copper phthalocyanine and poly(vinylcarbazole) are mixed[19] or a χ-form metal-free phthalocyanine

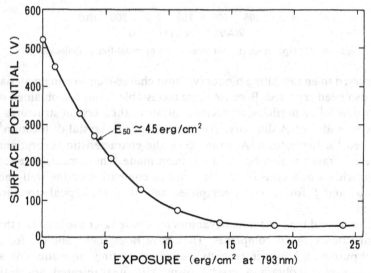

Figure 5. The photodecay curve for the τ-phthalocyanine photoreceptor. From this figure, the sensitivity E_{50}, that is, the half-decay exposure, is obtained from the curve at 793 nm and has the value of 4.5 erg cm^{-2}

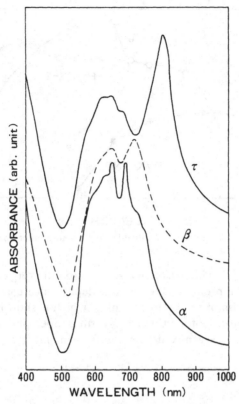

Figure 6. Absorption spectra for three forms of metal-free phthalocyanine.

is dispersed in an insulating binder (without charge-transporting material)[17] have also been reported. Because these metastable forms are obtained only in tiny needlelike particles, a precise analysis of their crystal structures has not been made yet. A dimerlike structure or specific crystal dislocation[25] is suggested for the χ-form. Assignment of the characteristic absorptions in the near-infrared region has also not been made. These metastable forms may provide a new class of crystal forms to compare with the well-understood α- and β-forms, but a complete analysis of the crystal structure is awaited.

The second type of phthalocyanines for diode-laser use includes three- or four-valence metal complexes. They have rarely been applied for pigmental purposes; however, recently they are attracting some attention as it appears easy to obtain a crystal form with near-infrared absorption. Aluminum, gallium, and indium phthalocyanine chlorides or bromides belong to the three-valence metal type,[23,26-28] and vanadyl (VOPc) or titanyl

(TiOPc) phthalocyanines,[29] the four-valence type. The vacuum-evaporated film of chloroaluminum phthalocyanine chloride, AlClPcCl, which shows only a visible spectrum, can be transformed by immersion in tetrahydrofuran or acetone to give another crystal form with near-infrared absorption.[30] This is achieved automatically when the CTL is coated from a solution onto the CGL. The dual-layered photoreceptor composed of the phthalocyanine and a pyrazoline derivative show a high sensitivity, about 6.3 erg cm^{-2}, in the 800–850-nm region. The crystal structure for one of the polymorphs (phase II) of vanadyl phthalocyanine has been determined as triclinic.[31,32] The form shows near-infrared absorption and can be applied to diode lasers,[33] where its maximum absorption is 815 nm. Similar polymorphism has been reported in titanyl phthalocyanine, and two crystal phases were analyzed, one of which is a similar triclinic phase (phase II) and the other (phase I) is monoclinic.[34] In these structures, the phthalocyanine ring is not planar, and the column structure, which is common in two-valence metal phthalocyanines or metal-free phthalocyanine, disappears. Presently, titanyl phthanocyanines, for which several different subphase forms are claimed, are the most widely applied in commercial-stage photoreceptors for diode-laser printers.[35,36] The spectral sensitivity of the dual-layered photoreceptor using α-titanyl phthalocyanine[35] is shown in Figure 7. Recently, a soluble

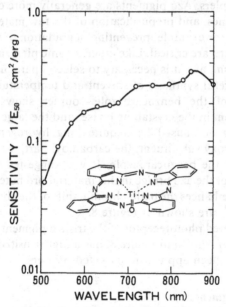

Figure 7. Spectral sensitivity for the dual-layered photoreceptor using α-titanyl phthalocyanine.

phthalocyanine with bulky substituents was developed to study the effect of aggregation of molecules on photoconductivity.[37,38]

B. Azo Pigments

Polyazo pigments are widely used as coloring agents for many purposes. One of the reasons why is the capability to modify their molecular structures by comparatively simple chemical syntheses. In electrophotographic applications, one disazo pigment, chlorodiane blue, was adopted as the charge-generating material in an early double-layered photoreceptor for copiers.[39,40] To enhance the photoconductivity and to extend the sensitivity range to longer wavelengths, many kinds of pigments have been obtained by modifying the center bridging part and/or the end parts with different coupling agents.[4,41-43] For the bridging part, several aromatic rings, including biphenyl, phenylbenzoxazole, diphenyloxadiazole, fluorenone, and styrylstilbene, have been applied. 2-Phenylcarbamoyl naphthol (Naphthol AS) and its derivatives have been used as the coupler, the same as in conventional azo pigments. To extend the sensitivity to longer wavelengths, trisazo type pigments have been developed.[44,45] Triaminotriphenylamine forms the bridging part and several phenylcarbamoylhydroxy derivatives of naphthalene, anthracene, dibenzofuran, carbazole, and benzo[a]carbazole acts as couplers. Azo pigments are generally more difficult to purify than phthalocyanines, and prepurification of the raw materials and a pure synthesis process, for example, preventing degradation of the azonium salt of the bridging parts, are critical. Like other organic pigments, azo pigments show polymorphism, and it is necessary to select a particular crystal structure, based on careful synthesis of solvent and temperature selection. The trisazo pigment of the benzocarbazole coupler shows an absorption maximum at 705 nm in the crystalline phase, and the large absorption shift was postulated to be caused by a quite long hydrogen-bonding chain involving the hydroxy substituent, the carbamoyl part, the azo bridge, and the amino part of the benzocarbazole. It was suggested that the ketohydrazone structure of the azo bridge in this material promotes an aggregation of molecules and enhances the bathochromic shift of the absorption. Typical polyazo pigments[45] are shown in Figure 8.

A double-layered photoreceptor of the trisazo pigment of the benzocarbazole-type coupler, shown in Figure 8, has a high sensitivity, 3.6 erg cm^{-2}, at 800 nm and has been applied in marketed printers.

C. Squarylium Pigments

Squarylium dyes or pigments (squaraines) were first examined for solar cell use and later applied to electrophotography.[39,46-49] They have strong

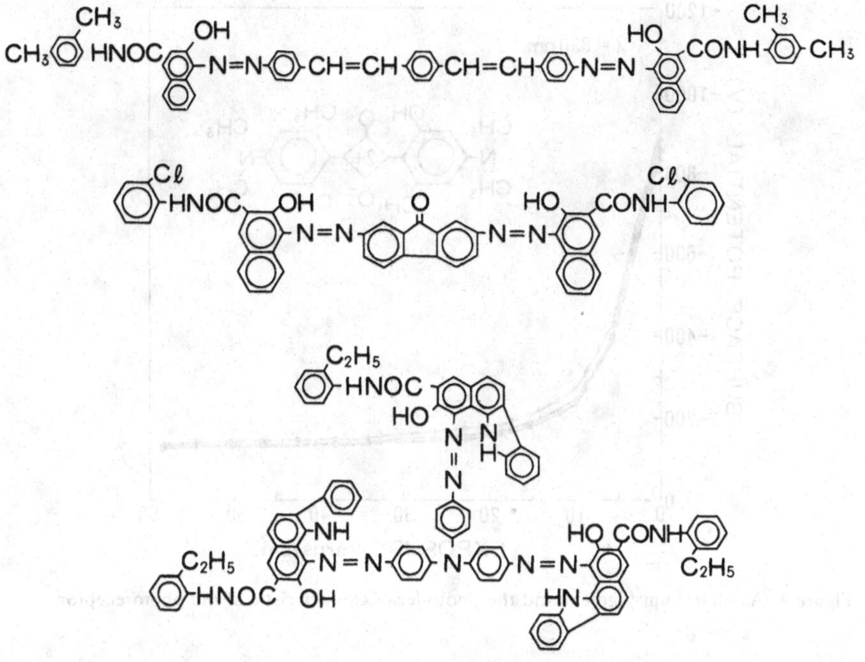

Figure 8. Typical polyazo pigments applied in photoreceptors.

intermolecular interactions and show characteristic absorptions in spite of rather small molecular structures. A polariton mechanism was suggested to explain the absorption characteristics of one squaraine, 2,4-bis(4-diethyl-amino-2-hydroxyphenyl)cyclobutadienediylium-1,3-diolate. Triclinic and monoclinic polymorphs of the squaraine were analyzed crystallographically. The triclinic phase shows absorptions in the near-infrared region. Several symmetrical and unsymmetrical squarines have been tested for photoreceptors. As shown in Figure 9, a symmetrical squaraine, bis(4-dimethylamino-2-hydroxy-6-methylphenyl)squaraine, has a photoconductivity of about 5 erg cm^{-2} at 830 nm.[49] The sensitivity is enhanced by addition of phenols, orcinol, or resorcinol during the condensation of squaric acid and 3-dimethylaminomethylphenol. The effect was ascertained to be caused by including different squaraine molecules in the crystalline framework; however, a precise explanation is still awaited.

D. Other Pigments

Perylene pigments are another important group of charge-generating materials for copier photoreceptors.[50-53] These are hard to dissolve in

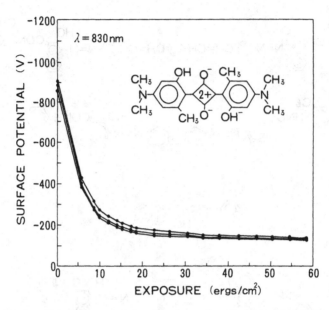

Figure 9. A squarylium pigment and the photodecay characteristic of the photoreceptor.

common solvents, and evaporated films are generally applied in practical manufacturing of CGLs. The deposited films can be used for CGL without any solvent or other treatments, unlike phthalocyanines, because the stable form (β) itself shows good photoconductivity.

Structural modifications have also been tried to extend the sensitivity range to longer wavelengths. The third compound in Figure 10 shows a spectral photosensitivity to the longest wavelength yet, 750 nm,[50] but that is a little too short for GaAlAs diode-laser application.

Another in this last group of photoreceptors is an aggregated photoreceptor.[54-57] This photoreceptor is composed of three components: **I**, a thiapyrylium salt [4-(4-dimethylaminophenyl)-2,6-diphenylthiapyrylium perchlorate]; **II**, a polycarbonate; and **III**, a triphenylmethane [bis(diethylaminomethylphenyl)phenylmethane], as shown in Figure 11. It was reported that their mixture spontaneously formed a complex system with a solvent treatment (e.g., dichloromethane), and the absorption spectra changed drastically to show a peak at 690 nm characteristic of an aggregated state and the photoconductivity increased to 4.1 erg cm^{-2} at 680 nm with negative charging. This photoreceptor also shows sensitivity for positive charging as shown in Figure 12, with a different spectral dependence. The crystal structure was determined for a model complex of **I** and a monomer of **II**, 4,4′-isopropylidenediphenyl-bis-phenylcarbonate. A two-dimensional

I

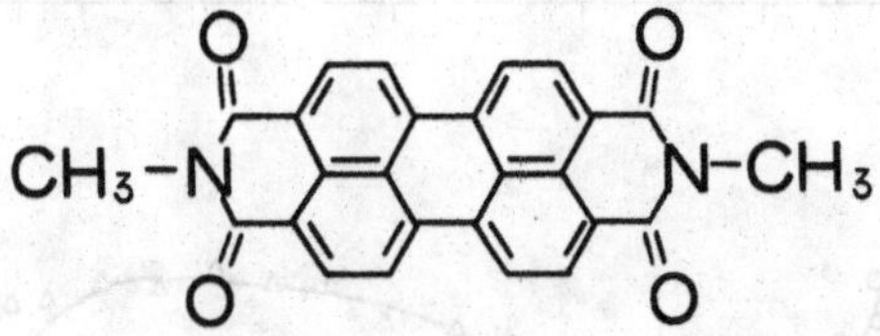

II

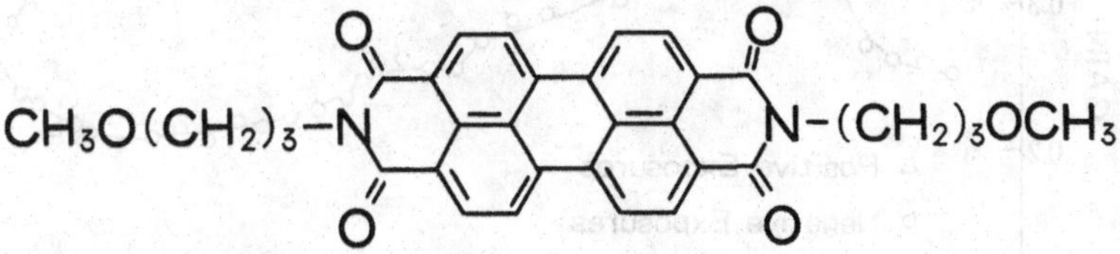

III

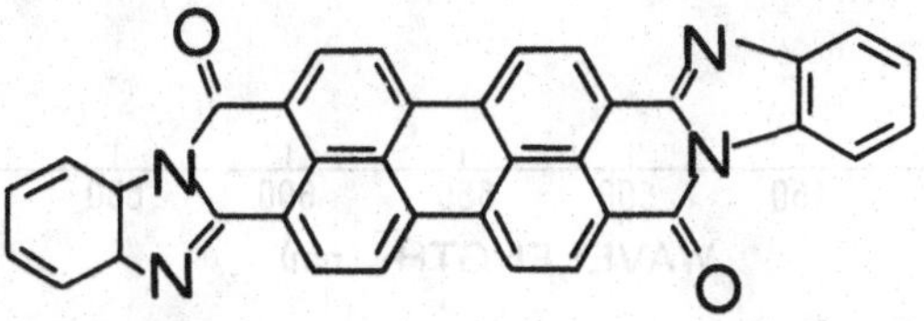

Figure 10. Typical perylene pigments.

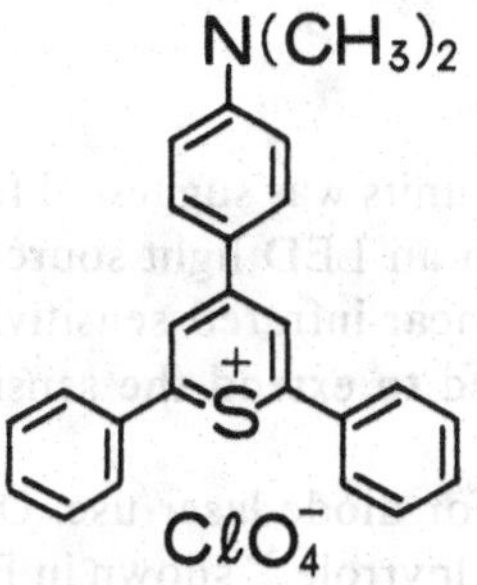

I

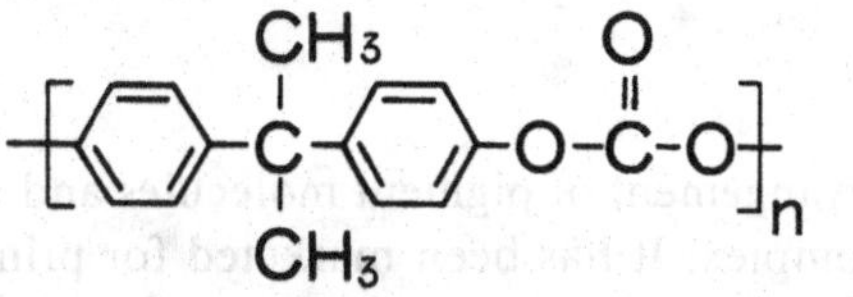

II

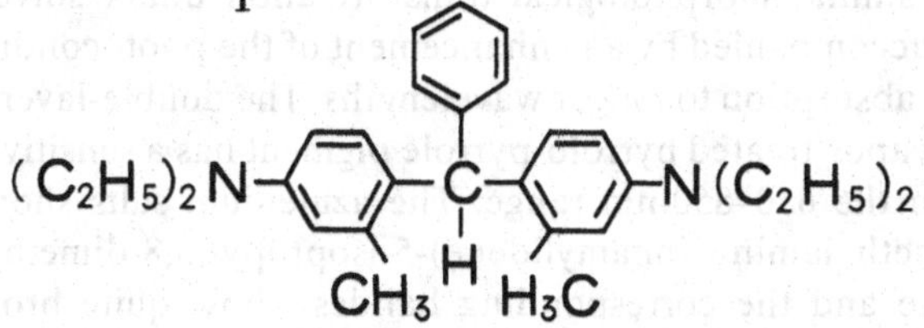

III

Figure 11. Chemical structures of the components of the aggregated photoreceptor.

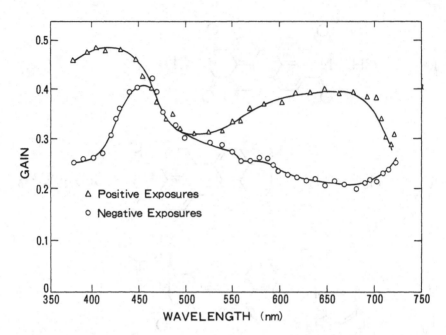

Figure 12. Spectral sensitivities of the aggregated photoreceptor for both negative and positive charging. The gains are not the same in the two cases. Refer to the text for details.

arrangement of pigment molecules and polymer units was suggested for the complex. It has been marketed for printers with an LED light source, but not for diode lasers because of its insufficient near-infrared sensitivity. A telluropyrylium-type pigment has been proposed to extend the sensitivity to longer wavelengths.

Other new pigments have been proposed for diode-laser use. One of them is 1,4-dithioketo-3,6-diphenylpyrrolo[3,4-c]pyrrole,[58] shown in Figure 13, which exhibits a similar morphological transformation under solvent (e.g., acetone) vapor accompanied by an enhancement of the photoconductivity and a shift of its absorption to longer wavelengths. The double-layered photoreceptor of the vapor-treated pyrrolo-pyrrole pigment has a sensitivity of 5.0 erg cm^{-2} within the 650–850-nm range. The azulenium salts shown in Figure 14, 1-(p-dimethylaminocinnamylidene)-5-isopropyl-3,8-dimethyl-azulenium perchlorate and the corresponding halides, show quite broad photosensitivity in the crystalline form, and when they are incorporated into a double-layered photoreceptor, a high sensitivity of $E_{50} = 2.5$ erg cm^{-2} is reached at 850 nm.[59]

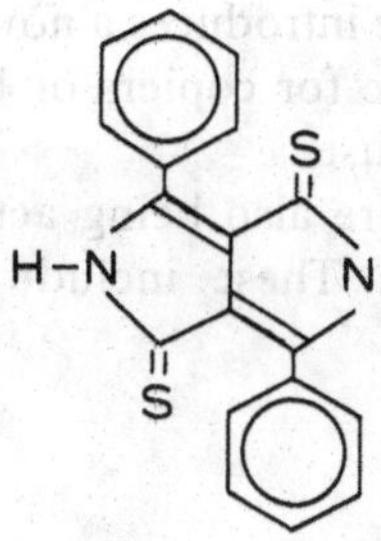

Figure 13. Pyrrolo-pyrrole pigment.

V. Conclusion

The present status of development and utilization of organic pigmental materials for diode-laser printers has been surveyed in this chapter. The materials can be divided into four groups: (1) phthalocyanines, (2) azo pigments, (3) squaraine pigments, and (4) perylene pigments, aggregated pigments, and others. The phthalocyanines are the oldest and still the most widely applied in practical photoreceptors. They are of two types—conventional metal-free or bivalent metal phthalocyanines transformed to a special crystal structure and tri- or tetravalent metal phthalocyanines having intrinsic near-infrared absorptions. The polyazo pigments are popular coloring agents and are commercially used as visible-sensitive photoreceptors for copiers. However, in order to apply them to diode lasers, bond conjugation in the molecular structures must be expanded and special trisazo pigments have to be designed and synthesized. These pigments are also applied in printers. Squaraine pigments are newcomers and have no counterpart among practical coloring agents. They also show quite strong absorbance in the near-infrared region and are being marketed. Perylene pigments are conventional coloring agents; however, because of their molecular structures, it is still hard to achieve a sufficiently strong near-infrared

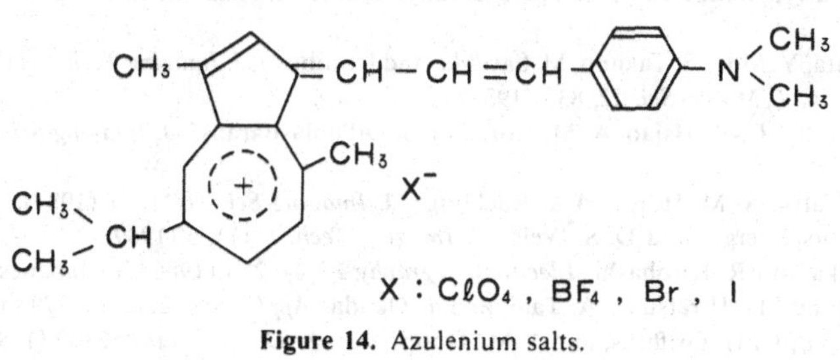

X : ClO_4 , BF_4 , Br , I

Figure 14. Azulenium salts.

absorption. The aggregated type introduces a novel aspect of organic photo-receptors. This type is available for copiers or LED printers, but it needs to be modified for diode-laser use.

Other types of pigments are also being actively developed and their application in printers tested. These include pyrrolo-pyrrole types or azulenium types.

REFERENCES

1. R. M. Schaffert, *Electrophotography*, The Focal Press, London (1975).
2. R. M. Schaffert, *IBM J. Res. Develop. 15*, 75 (1971).
3. A. Kakuta, Y. Mori, and H. Morishita, *IEEE Trans. Ind. Appl. IA-17*(4), 382 (1981).
4. M. Umeda, *Electrophotography 27*(4), 539 (1988) (in Japanese).
5. J. E. Kuder, J. M. Pochan, S. R. Turner, and D. F. Hinman, *J. Electrochem. Soc. 125*(11), 1750 (1978).
6. B. S. Ong, B. Keoshkerian, T. I. Martin, and G. K. Hamer, *Can. J. Chem. 63*, 147 (1985).
7. M. Stolka, J. F. Yanus, and D. M. Pai, *J. Phys. Chem. 88*, 4707 (1984).
8. T. Kitamura and M. Yokoyama, *Electrophotography 27*(3), 406 (1988) (in Japanese).
9. D. M. Pai, *J. Non-Cryst. Solids 59 & 60*, 1255 (1983).
10. R. Takahashi, S. Kusabayashi, and M. Yokoyama, *Electrophotography 25*(3), 236 (1986) (in Japanese).
11. T. Kitamura and M. Yokoyama, *Electrophotography 27*(1), 31 (1988) (in Japanese).
12. Y. Kanemitsu and S. Imamura, *Solid State Commun. 63*, 1161 (1987).
13. Y. Kanemitsu and S. Imamura, *J. Appl. Phys. 63*, 239 (1988).
14. E. S. Baltazzi, *J. Imaging Tech. 10*(6), 120A (1984).
15. E. S. Baltazzi, *J. Imaging Tech. 12*(2), 9A (1986).
16. A. B. P. Lever, *Adv. Inorg. Chem. Radiochem. 7*, 27 (1965).
17. J. W. Weigl, J. Mammino, G. L. Whittaker, R. W. Radler, and J. F. Byrne, in: *Current Problems in Electrophotography* (W. F. Berg and K. Hauffe, eds.), p. 287, Walter de Gruyter, Berlin (1972).
18. C. F. Hackett, *J. Chem. Phys. 53*, 3178 (1971).
19. T. Yagishita, K. Ikegami, T. Narusawa, and H. Okuyama, *IEEE Trans. Ind. Appl. IA-20*, 1642 (1984).
20. S. Takano, T. Enokida, A. Kakuta, and Y. Mori, *Chem. Lett.* 2037 (1984).
21. T. L. Bluhm, H. J. Wagner, and R. O. Loutfy, *J. Mater. Sci. Lett. 2*, 85 (1983).
22. T. Kobayashi, N. Uyeda, and E. Suito, *J. Phys. Chem. 72*, 2446 (1968).
23. R. O. Loutfy, A. M. Hor, G. DiPaola-Baranyi, and C. K. Hsiao, *J. Imaging Sci. 29*(3), 116 (1985).
24. A. Kakuta, Y. Mori, S. Takano, M. Sawada, and I. Shibuya, *J. Imaging Tech. 11*(1), 7 (1985).
25. A. Perovic, *J. Mater. Sci. 22*, 835 (1987).
26. R. O. Loutfy, C.-K. Hsiao, A. M. Hor, and G. DiPaola-Baranyi, *J. Imaging Sci. 29*(4), 148 (1985).
27. R. O. Loutfy, A. M. Hor, and A. Rucklidge, *J. Imaging Sci. 31*(1), 31 (1987).
28. P. M. Borsenberger and D. S. Weiss, *J. Imaging Tech. 15*(1), 6 (1989).
29. T. Tanaka and R. Hirohashi, *Electrophotography 25*(4), 284 (1986) (in Japanese).
30. K. Arishima, H. Hiratsuka, A. Tate, and T. Okada, *Appl. Phys. Lett. 40*, 279 (1982).
31. R. F. Ziolo, C. H. Griffiths, and J. M. Troup, *J. C. S. Dalton Trans.*, 2300 (1980).

32. C. H. Griffiths, M. S. Walker, and P. Goldstein, *Mol. Cryst. Liq. Cryst. 33*, 149 (1976).

33. S. Grammatica and J. Mort, *Appl. Phys. Lett. 38*, 445 (1981).

34. W. Hiller, J. Strahle, W. Kobel, and M. Hanack, *Z. Kristallogr. 159*, 173 (1982).

35. K. Ohaku, H. Nakano, T. Kawara, S. Yokoyama, O. Takenouchi, and M. Aizawa, *Electrophotography 25*(3), 258 (1986) (in Japanese).

36. T. Enokida, R. Kurata, T. Seta, and H. Katsura, *Electrophotography 27*(4), 533 (1988) (in Japanese).

37. K.-Y. Law, *J. Phys. Chem. 89*, 2652 (1985).

38. K.-Y. Law, *J. Phys. Chem. 92*, 4226 (1988).

39. P. J. Melz, R. B. Champ, L. S. Chiou, G. S. Keller, L. C. Liclican, R. R. Neiman, M. D. Shattuck, and W. J. Weiche, *Photogr. Sci. Eng. 21*, 73 (1977).

40. N. C. Khe, O. Takanouchi, T. Kawara, H. Tanaka, and S. Yokota, *Photogr. Sci. Eng. 28*, 195 (1984).

41. T. Kazami, Ricoh Technical Report No. 3, p. 4 (May, 1980) (in Japanese).

42. M. Sasaki, *Nippon Kagaku Kaishi*, 379 (1986) (in Japanese).

43. M. Hashimoto, *Electrophotography 25*(3), 230 (1986) (in Japanese).

44. M. Ota, Ricoh Technical Report No. 8, p. 14 (November, 1982) (in Japanese).

45. K. Ohta, *Electrophotography 25*(3), 303 (1986) (in Japanese).

46. K.-Y. Law and F. C. Bailey, *J. Imaging Sci. 31*(4), 172 (1987).

47. M. Tristani-Kendra and C. J. Eckhardt, *J. Chem. Phys. 81*, 1160 (1984).

48. P. M. Kazmaier, R. Burt, G. DiPaola-Baranyi, C.-K. Hsiao, R. O. Loutfy, T. I. Martin, G. K. Hamer, T. L. Bluhm, and M. G. Taylor, *J. Imaging Sci. 32*(1), 1 (1988).

49. G. DiPaola-Baranyi, C.-K. Hsiao, P. M. Kazmaier, R. Burt, R. O. Loutfy, and T. I. Martin, *J. Imaging Sci. 32*(2), 60 (1988).

50. D. Winkelman and K. Reuter, 1st International Congress on Advances in Non-impact Printing Technologies, Venice (June 1981).

51. Z. D. Popovic, R. O. Loutfy, and A. M. Hor, *Can. J. Chem. 63*, 134 (1985).

52. N. C. Khe, S. Yokota, and K. Takahashi, *Photogr. Sci. Eng. 28*, 191 (1984).

53. E. G. Schlosser, *J. Appl. Photogr. Eng. 4*(3), 118 (1978).

54. W. Wey, E. I. P. Walker, and D. C. Hoesterey, *J. Appl. Phys. 50*, 8090 (1979).

55. P. M. Borsenberger and D. C. Hoesterey, *J. Appl. Phys. 51*, 4248 (1980).

56. W. J. Dulmage, W. A. Light, S. J. Marino, C. D. Salzberg, D. L. Smith, and W. J. Staudenmayer, *J. Appl. Phys. 49*, 5543 (1978).

57. P. M. Borsenberger, A. Chowdry, D. C. Hoesterey, and W. Mey, *J. Appl. Phys. 49*, 5555 (1978).

58. J. Mizuguchi and A. C. Rochat, *J. Imaging Sci. 32*(3), 135 (1988).

59. K. Katagiri, Y. Oguchi, and Y. Takasu, *Nippon Kagaku Kaishi*, 387 (1986) (in Japanese).

13
Laser Filter Systems

YOSHIAKI SUZUKI

I. INTRODUCTION

Lasers are sources of light possessing properties which differ from those of light from conventional sources. Laser light can be highly monochromatic, very well collimated, and coherent, and in some cases lasers are extremely powerful. These characteristics make the laser a very useful source of light for a variety of applications in science and industry. Laser wavelengths cover the far-infrared to the near-infrared, the visible to the ultraviolet, and the vacuum ultraviolet to the soft X-ray region.

Optical filters are used widely in engineering and research when the spectral distribution of the incident laser energy must be selectively or nonselectively altered or precisely controlled.

It is neither possible nor desirable to discuss in this chapter all the filters that are used for the many different types of lasers. Here the discussion is largely confined to absorption filters for lasers that produce light at wavelengths in the infrared spectrum.

The most common types of absorption filters are made of glass, gelatin, or plastics in which coloring agents are dissolved or suspended. The main emphasis in this chapter is on plastic filters. For glass filters, the reader is referred to the literature.[1,2]

YOSHIAKI SUZUKI • Late of Ashigara Research Laboratories, Fuji Photo Film Co., Ltd., Minami-Ashigara, Kanagawa 250-01, Japan.

II. GELATIN FILTERS

Gelatin filters are made by mixing water-soluble infrared absorbing dyes in gelatin and coating the mixture on glass plates. After the coatings are dried, the gelatin films are removed from the plates. After stripping from the plates, the gelatin film is either lacquered or cemented between glass plates for protection. However, many gelatin filters are not very stable with regard to prolonged exposure to radiant energy, high humidity, and temperature. Gelatin filters are relatively inexpensive but must be protected from moisture and cannot be used above 55°C. Since their transmission characteristics may change on exposure to light, they should be stored in a dark, dry place.

III. PLASTIC FILTERS

Plastic filters are very inexpensive and generally stable. They can be cut readily to the desired size and shape and are flexible and convenient to use. Although such filters are generally less stable than glass filters, their spectral features are often much sharper.

Plastic filters consist of infrared absorbing dyes dissolved in some transparent host medium in amounts such as to provide the desired optical properties. The plastic substrate used in the preparation of the filters is not critical so long as it meets the requirements of the resulting filter, that is, is self-supporting, durable, stable, easily manufactured, and so on, and does not possess optical properties which interfere with those of the filter. The plastics used include cellulose acetate, polypropylene, poly(ethylene terephthalate), poly(methyl methacrylate), polystyrene, and polycarbonate.

Plastics fall into two categories, thermoplastics and thermosets, depending on how heat affects them. Almost without exception, filters made from thermoplastics are prepared by melting thermoplastic resins and dyes, shaping the melt, and cooling it while maintaining the shape of filters [e.g., poly(methyl methacrylate)]. When the plastics are not thermally stable, appropriate low-boiling organic solvents are often used (e.g., cellulose triacetate). In the case of thermosets, the polymerization reaction is only partially complete when the dyes are added and is concluded when in the molding stage (e.g., epoxy resins).

A typical example of the preparation of a plastic filter is as follows. The components shown below are mixed and stirred well and the mixture is filtered. The solution is casted on a metal support and allowed to dry to form a uniform film. After evaporation of the solvents, the film is stripped

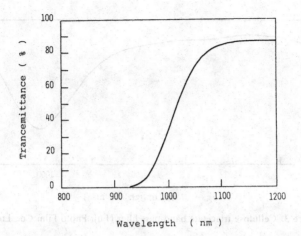

Figure 1. Cellulose triacetate cutoff filter (Fuji Photo Film Co., Ltd.).

off to provide a film having a thickness of 25 μm.

Composition of Film	
Cellulose triacetate	170 parts
Triphenyl phosphate	10 parts
Methylene chloride	800 parts
Methanol	160 parts
Infrared absorbing dye	2 parts

Some examples of practical filters are shown in Figures 1–3.

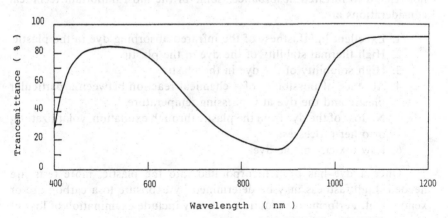

Figure 2. Cellulose triacetate band-pass filter (Fuji Photo Film Co., Ltd.).

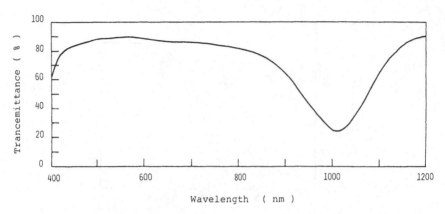

Figure 3. Cellulose triacetate band-pass filter (Fuji Photo Film Co., Ltd.).

IV. INFRARED ABSORBING DYES AND PLASTICS

Of course, not every material which exhibits spectrally selective absorption is a potential filter material; another important requirement is that its optical behavior be stable with time and insensitive to reasonable fluctuations in the operating conditions.

A. Selection of Dyes

The choice of an infrared absorbing dye for a particular plastic and use is governed largely by economic factors and by technical considerations not related to infrared absorbance. Some of the most important technical considerations are:

1. Excellent lightfastness of the infrared absorbing dye in the plastic.
2. High thermal stability of the dye in the plastic.
3. High solubility of the dye in the plastic.
4. Absence of possibility of a chemical reaction between a particular plastic and the dye at processing temperature.
5. No loss of the dye from the plastic through exudation, volatilization, or other processes.
6. Low toxicity of the dye.

Once a dye has been incorporated into the plastic, more tests are needed. Lightfastness may be determined by exposure to a carbon arc or xenon light. Performance requirements may include examination of loss of tensile strength or change in brittleness.

B. Classification of Infrared Absorbing Dyes

Infrared absorbing dyes for filters may be classified into two groups: organic dyes and metal complexes.

1. Organic Dyes

Generally, the color of an organic dye originates from electronic transitions in the molecule.[3,4] A bathochromic shift is produced by expanding the conjugation system of the molecule. Because of poor fastness properties, especially to light, many organic infrared absorbing dyes are not practically used as materials for laser filters.

2. Metal Complexes

Metal complexes are generally very stable to light and heat and are very important as dyes for laser filters in the practical sense.

The central metal coordinated to the organic ligand in these complexes is usually a transition metal. A transition metal complex undergoes electron transitions by the absorption of light in the ultraviolet, visible, and near-infrared regions.[5] Not all electron transitions are allowed by quantum mechanics. The two principal selection rules, the spin selection rule and Laporte's selection rule, are involved in establishing the conditions under which certain electron transitions may occur.[6-8]

The electronic spectra of transition metal complexes mainly arise from d–d transitions, charge-transfer transitions, and internal ligand transitions. Electron transitions between MOs mainly localized on the central metal are usually called d–d transitions. Although d–d transitions are spin-forbidden and Laporte forbidden, there is always some low-intensity absorption because of the weak coupling of spin and orbital angular momenta and the mixing to some extent of the metal d orbitals with p or f orbitals. They have molar absorptivities of the order of 10^{-3}–10^{-2} dm^3 mol^{-1} cm^{-1}. These bands are weak and not very significant in the application of these complexes as laser filter dyes.

Transitions between MOs mainly localized around the ligand and MOs mainly localized on the central metal are called charge-transfer transitions. Charge-transfer bands are generally very intense, with molar absorptivities of the order of 10^4 dm^3 mol^{-1} cm^{-1}, representing nearly full spin-allows and Laporte allowed transitions.

With dithiolene complexes,[9] these charge-transfer absorption bands often shift into the infrared region.

Table I. Infrared Absorbing Dyes (Organic Dyes)

Class	Basic chemical structure	Ref.
Cyanine dyes		19,20
Pyrylium and thiapyrylium dyes		21, 22
Squarylium dyes		23, 24
Indoaniline dyes		25,26
Azo dyes		3, 27
Anthraquinone dyes		3, 28
Naphthoquinone dyes		29, 30

Table I. (continued)

Class	Basic chemical structure	Ref.
Aminium radical salts	$\left[Ar_3N^{\cdot+}\right] X^-$	31, 32
Charge-transfer complexes	$\left[Ar_2N-\langle\text{benzene}\rangle-NAr_2\right]\left[TCNQ\right]$	33,34

Transitions between MOs mainly localized on the ligands are called internal ligand transitions. These transitions only involve ligand orbitals that are almost unaffected by coordination to the central metal. These transitions are spin-allowed and Laporte allowed and have molar absorptivities of the order of 10^3-10^5 dm^3 mol^{-1} cm^{-1}. Such complexes as metalated azo dyes and metal phthalocyanines have absorptions of very high intensity for this reason.[10-16]

The lightfastness of dyes such as azo dyes and phthalocyanines may also be improved by metalation, which causes a perturbation of the π-electron distribution of the molecule. The absorption maximum often shifts to longer wavelength. The recent synthesis of infrared absorbing metalated indoaniline-type dyes[17] is of much interest. These can also be expected to have good fastness.

Metal phthalocyanines show characteristic absorption spectra consisting of four intense $\pi-\pi^*$ bands in the uv–vis region and are remarkably stable to heat, light, and chemical reagents. Recently, active research on near-infrared absorbing metal phthalocyanines which are very soluble in low-boiling solvents has been undertaken.

Infrared absorbing dyes for laser filters found in the literature may be classified based on chemical structure as shown in Tables I and II.

V. FILTER APPLICATION

Laser filters are now being widely used in electronics research and the electronics industry. There is another interesting application of laser filters. Laser radiation, especially from high-power lasers, is employed for a variety of material processing functions, such as welding, cutting, shaping, and drilling, in industry and for bloodless surgery. The beam from a laser often presents a serious hazard to the human eye, necessitating eye protection equipment.[18] Laser filters can be used as safety glasses, making possible unobstructed yet safe viewing and working.

Table II. Infrared Absorbing Dyes (Metal Complexes)

Class	Basic chemical structure	References
Metal salts	MX_n (M = Cu, W)	35–37
Metalated azo dyes		38, 39
Metal phthalocyanines		11, 40
Bis(dithiolene) complexes I		9, 41, 42
Bis(dithiolene) complexes II		43
α-Diimine-dithiolene complexes		44–46
Tris(α-diimine) complexes		47, 48

ACKNOWLEDGMENTS

I would like to acknowledge Goichi Hayashi of Ashigara Factory, Fuji Photo Film Co., Ltd., for permission to reproduce his spectra. Special thanks to Hiromichi Tateishi of Ashigara Research Laboratories, Fuji Photo Film Co., Ltd., for assistance in the preparation of the references for this chapter.

REFERENCES

1. E. K. Letzer, in: *Kirk-Othmer Encyclopedia of Chemical Technology*, 3rd ed. (M. Grayson, ed.), Vol. 16, p. 522, John Wiley and Sons, New York (1981).
2. J. A. Dobrowolski, in: *Handbook of Optics* (W. G. Driscoll, ed.), p. 81, McGraw-Hill, New York (1978).
3. P. F. Gordon and P. Gregory, *Organic Chemistry in Colour*, Springer-Verlag, Berlin (1983).
4. H. Zollinger, *Colour Chemistry*, VCH, Weinheim (1987).
5. A. B. P. Lever, *Inorganic Electronic Spectroscopy*, 2nd ed., Elsevier, Amsterdam (1984).
6. T. M. Dunn, in: *Modern Coordination Chemistry* (J. Lewis and R. G. Wilkins, eds.), p. 229, Interscience, New York (1960).
7. V. Balzani and V. Carassiti, *Photochemistry of Coordination Compounds*, Academic Press, London (1970).
8. A. W. Adamson and P. D. Fleishauer, *Concepts of Inorganic Photochemistry*, John Wiley and Sons, New York (1975).
9. G. N. Schrauzer, *Acc. Chem. Res.* 2, 72 (1969).
10. A. B. P. Lever, in: *Advances in Inorganic Chemistry and Radiochemistry* (H. J. Eméleus and A. G. Sharpe, eds.), Vol. 7, p. 27, Academic Press, New York (1965).
11. F. H. Moser and A. K. Thomas, *The Phthalocyanines*, Vols. 1 and 2, CRC Press, Boca Raton, Florida (1983).
12. K. M. Smith, in: *Comprehensive Heterocyclic Chemistry* (A. R. Katritzky and C. W. Rees, eds.), Vol. 4, p. 377, Pergamon Press, Oxford (1984).
13. K. Venkataraman, *The Chemistry of Synthetic Dyes*, Vol. I, Academic Press, New York (1952).
14. R. D. Johnson and N. C. Nielsen, in: *The Chemistry of the Coordination Compounds* (J. C. Bailar, Jr., ed.), p. 743, Reinhold, New York (1956).
15. H. Baumann and H. R. Hensel, *Fortschr. Chem. Forsch.* 7, 643 (1967).
16. R. Price, in: *The Chemistry of Synthetic Dyes* (K. Venkataraman, ed.), Vol. III, p. 303, Academic Press, New York (1970).
17. Y. Kubo, K. Sasaki, and K. Yoshida, *Chem. Lett.*, 1563 (1987).
18. D. J. Spencer and H. A. Bixler, *Rev. Sci. Instrum.* 43, 1543 (1972).
19. G. E. Ficken, in: *The Chemistry of Synthetic Dyes* (K. Venkataraman, ed.), Vol. IV, p. 211, Interscience, New York (1971).
20. G. W. Ingle and W. B. Teummler, U.S. Patent 2,813,802.
21. F. H. Hamer, in: *The Chemistry of Heterocyclic Compounds* (A. Weissberger, ed.), Vol. 18, p. 495, Interscience, New York (1964).
22. K. Katagiri and Y. Takasu, Jpn. Patent 58 217,558 [*CA 100*, 193532 (1983)].
23. K. Y. Law and F. C. Baily, *Dyes Pigments 9*, 85 (1988).
24. K. Miura, J. Iwanami, and T. Ozawa, Jpn. Patent 61 218,551 [*CA 106*, 129460 (1986)].

25. K. Venkataraman, *The Chemistry of Synthetic Dyes*, Vol. II, p. 762, Academic Press, New York (1952).
26. T. Kitao and M. Matsuoka, Jpn. Patent 62 201,857.
27. S. Kawasaki, H. Nishii, and H. Hino, Jpn. Patent 61 263,958 [*CA 106*, 175939 (1986)].
28. K. Kojima, Jpn. Patent 62 903 [*CA 106*, 204968 (1986)].
29. T. Kitao and M. Matsuoka, Jpn. Patent 62 10,092 [*CA 107*, 156367 (1987)].
30. T. Ozawa, S. Maeda, and Y. Kurose, Jpn. Patent 62 56,460 [*CA 107*, 124746 (1987)].
31. A. E. Sherr, R. J. Tucker, R. H. Spector, and H. G. Brooks, Jr., U.S. National Technical Information Service, AD-D Report, No. 006575 [*CA 83*, 157087 (1975)].
32. R. A. Coleman and P. V. Susi, U.S. Patent 3,341,464 [*CA 62*, 1816 (1965)].
33. M. Hiraishi and T. Ishihara, Jpn. Patent 58 99,446 [*CA 99*, 166773 (1983)].
34. M. Matsuoka, L. Han, T. Kitao, S. Mochizuki, and K. Nakatsu, *Chem. Lett.*, 905 (1988).
35. R. J. Hovey, U.S. Patent 3,629,130.
36. G. A. Caterion and J. P. Habermann, U.S. Patent 3,692,688.
37. N. U. Laliberte, U.S. Patent 3,826,751.
38. S. Kawasaki, H. Nishii, and H. Hino, Jpn. Patent 61 15,891 [*CA 105*, 42512 (1986)].
39. I. Niimura, S. Matsumoto, H. Yamaga, and S. Suzuka, Jpn. Patent 63 4,992.
40. T. Eda, U.S. Patent 4,622,179; Eur. Patent Appl. 134518 [*CA 103*, 38703 (1985)].
41. S. M. Bloom, U.S. Patent 3,588,216 [*CA 70*, 24553 (1969)].
42. Y. Suzuki and G. Hayashi, Jpn. Patent 61 26,686 [*CA 105*, 200221 (1986)].
43. S. M. Bloom, U.S. Patent 3,806,462 [*CA 81*, 8318 (1974)].
44. Y. Suzuki and G. Hayashi, Jpn. Patent 62 39,682 [*CA 107*, 22570 (1987)].
45. Y. Suzuki, Jpn. Patent 63 126,889.
46. Y. Suzuki and G. Hayashi, Jpn. Patent 63 139,303.
47. Y. Suzuki and G. Hayashi, Jpn. 61 20,002 [*CA 105*, 135071 (1986)].
48. Y. Suzuki and G. Hayashi, Jpn. Patent 61 73,902 [*CA 106*, 11097 (1986)].

14

Infrared Photography

TADAAKI TANI and YUJI MIHARA

I. INTRODUCTION

Photosensitive elements of photographic materials are silver halide microcrystals. As shown in Figure 1, there are many mobile interstitial silver ions in the interior, and there are sensitivity centers composed of silver gold sulfide on the surface of each microcrystal. The absorption of a photon by the microcrystal results in the formation of a free electron, which is soon trapped by a sensitivity center. Then, an interstitial silver ion reaches and reacts with the trapped electron to form a silver atom. The repetition of these electronic and ionic processes at the same center leads to the formation of a latent image, which is a cluster composed of several silver atoms and acts as the catalyst for the electrochemical reduction of the host microcrystal to form a silver or dye image during photographic development.[1]

However, spectral sensitivity of silver halide microcrystals is limited to ultraviolet and blue light. A technology named "spectral sensitization" is used to make photographic materials sensitive to green, red, and infrared light.[2] Namely, sensitizing dyes, which are absorbed on the surface of the silver halide microcrystals, inject electrons into the conduction band of the silver halide when they absorb light. Accordingly, for the preparation of infrared photographic materials, we need infrared sensitizing dyes, which can inject electrons into the silver halide by absorbing infrared light.

TADAAKI TANI and YUJI MIHARA • Ashigara Research Laboratories, Fuji Photo Film Co., Ltd., Minami-Ashigara, Kanagawa 250-01, Japan.

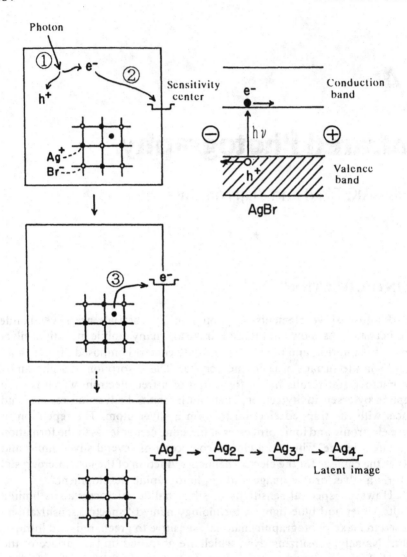

Figure 1. Mechanism of photographic sensitivity.

In the following sections, brief descriptions are given of infrared sensitizing dyes, spectral sensitization of silver halide microcrystals by infrared sensitizing dyes, and materials for infrared photography.

II. INFRARED SENSITIZING DYES

Since the discovery of spectral sensitization by Vogel in 1873,[3] the portion of the spectrum which can be photographed has been gradually extended to longer wavelengths. It was extended to beyond 700 nm with pinacyanol (1), which was invented in 1904 by Homolka,[4]

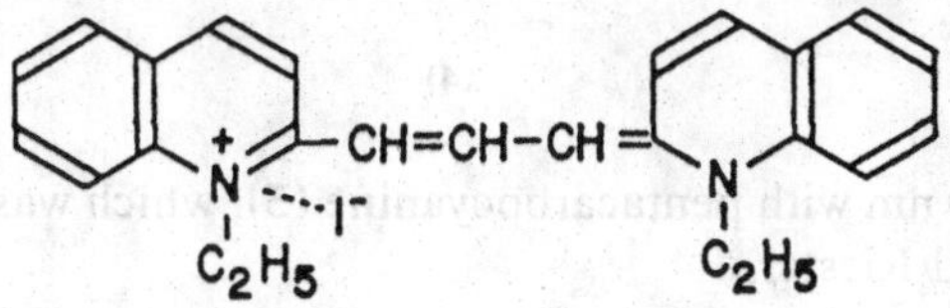

(1)

to 800 nm with kryptocyanine (2), which was invented in 1920 by Adams and Haller,[5]

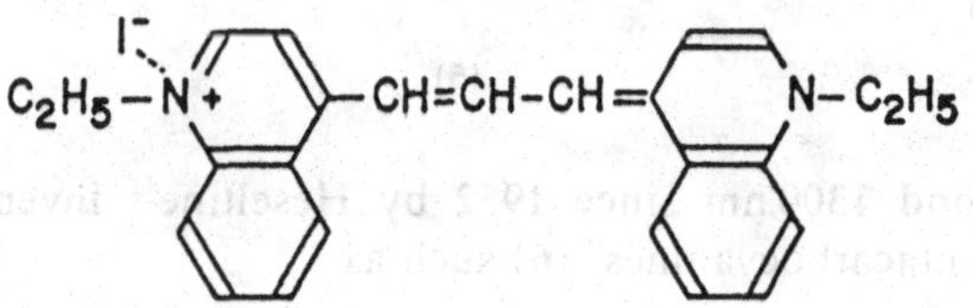

(2)

to 900 nm with neocyanine (3), which was invented in 1931 by Clarke,[6]

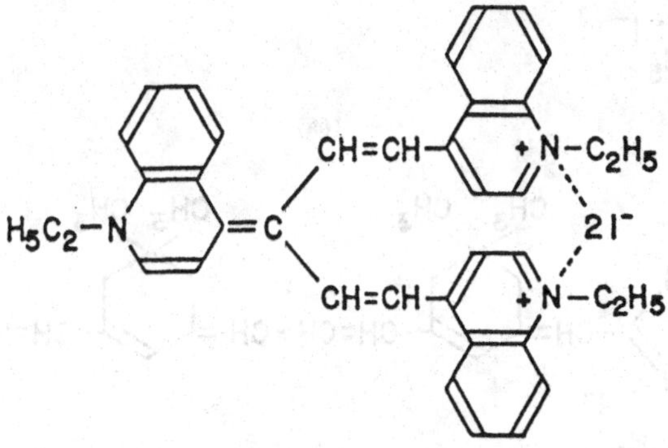

(3)

to 1100 nm with xenocyanine (4), which was invented in 1933 by Brooker et al.[7]

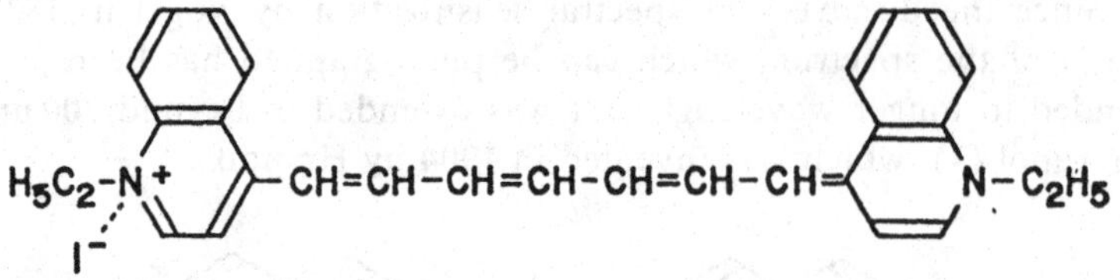

(4)

to beyond 1200 nm with pentacarbocyanine (5), which was invented in 1937 by Dieterle and Riester,[8]

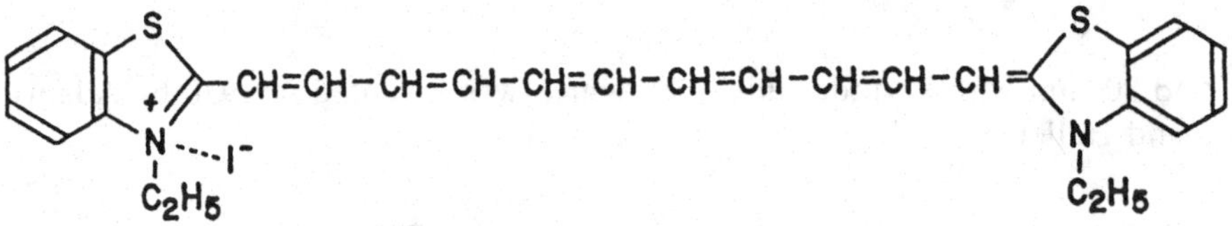

(5)

and to beyond 1300 nm since 1952 by Heseltine's invention of chain-stabilized pentacarbocyanines[9] (6) such as

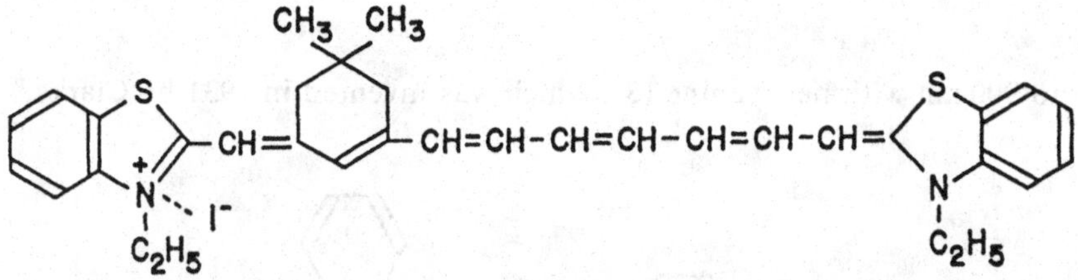

(6a)

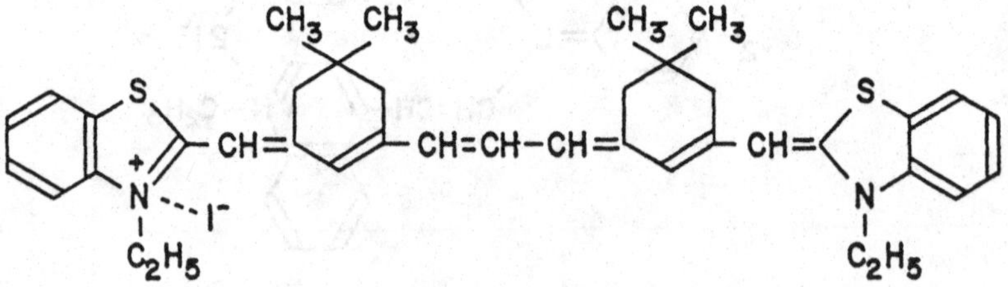

(6b)

Table I. Properties and Behavior of Thiacyanine Dyes with Variation of Polymethine Chain Length[a]

Dye	n	λ_{max}^{EtOH} (nm)	λ_{max}^{S} (nm)	ϕ_r	E_R^{b} (V vs. SCE)
a	0	425	450	0.74[c]	1.52
b	1	560	600	0.11,[c] 0.27[d]	1.13
c	2	655	700	0.02[d]	0.94
d	3	760	810[d]	0.02[d]	0.80
e	4	870	940		
f	5	995	1000		

[a] Definition of symbols: λ_{max}^{EtOH}, wavelength of absorption peak of the corresponding dye in ethanol; λ_{max}^{S}, wavelength of sensitivity peak of AgBr microcrystals sensitized by the corresponding dye; ϕ_r, relative quantum yield of spectral sensitization by the corresponding dye; E_R, polarographic reduction half-wave potential.
[b] Ref. 10.
[c] Ref. 11.
[d] W. West, *Photogr. Sci. Eng.* 6, 92 (1962).

The applicability of infrared sensitizing dyes depends upon their stability on storage, ability to absorb infrared light, and efficiency of spectral sensitization.

Most sensitizing dyes, including infrared ones, are of the cyanine or merocyanine type. Dyes of both types have a conjugated polymethine chain in their molecular structure.[10] An increase in the length of the polymethine chain in these dyes produces a bathochromic shift of λ_{max}, decreases their stability on storage, and decreases their efficiency of spectral sensitization, ϕ_r,[11] as shown in Table I. Thus, infrared sensitizing dyes are generally characterized by the presence of long polymethine chains in their molecular structure, low stability on storage, and low efficiency of spectral sensitization.

Extensive efforts have been made to alleviate the above problems associated with infrared sensitizing dyes. Introduction of a cyclic methine chain is very effective in improving the stability of infrared sensitizing dyes, as exemplified by dyes **6**. It is also important in the design of the molecular structure of infrared sensitizing dyes to consider their efficiency of spectral sensitization, which is described in the next section.

III. SPECTRAL SENSITIZATION BY INFRARED SENSITIZING DYES

Spectral sensitization in silver halide photography is initiated by light absorption by a sensitizing dye molecule, as seen in Figure 2. The light absorption results in the transition of an electron from the highest occupied molecular orbital (HOMO) to the lowest unoccupied one (LUMO), and electron transfer from the excited dye molecules to the conduction band of silver halide follows. Then, elucidation of the energy gap (ΔE) dependence of the efficiency of spectral sensitization (ϕ_r) is essential for the understanding of the spectral sensitization in photography. Reduction potentials E_R of the dyes could serve as a measure of the energy level of LUMO,[2,10] and therefore of the value of the energy gap.

The values of ϕ_r, and its temperature coefficient E_a for various cyanine dyes on AgBr microcrystals are plotted against the values of E_R of the corresponding dyes in Figure 3.[11] It is noticed that a value of -1.2 V for E_R was the threshold for the electron transfer. Namely, the dyes whose E_R is more negative than -1.2 V have large and temperature-independent ϕ_r, whereas the values of ϕ_r for dyes with E_R less negative than -1.2 V are low and temperature-dependent.

The values of E_R and of the wavelength of the absorption peak on AgBr microcrystals of thiacyanine dyes with variation of the polymethine chain length are listed in Table I. As shown in the table, the energy level of the LUMO of the dyes, as judged from their E_R, decreases with increase

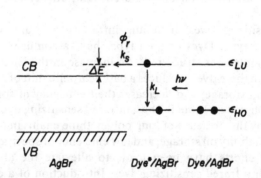

Figure 2. Schematic representation showing electron-transfer process in spectral sensitization in photographic materials, where ε_{LU} and ε_{HO} are the electronic energy levels of the lowest unoccupied molecular orbital (LUMO) and the highest occupied one (HOMO), respectively, k_s and K_l are the rate constants of the electron transfer and other processes, respectively, $\phi_r = k_s/(k_s + k_l)$ is the efficiency, and ΔE is the energy gap of the electron-transfer process.

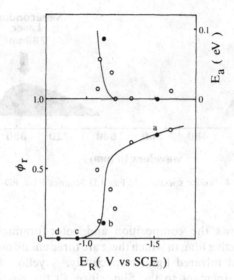

Figure 3. Relative quantum yield of spectral sensitization (ϕ_r) and its temperature coefficient (E_a) on AgBr microcrystals of various cyanine dyes including dyes a, b, c, and d in Table I as a function of their reduction potential E_R.

in the chain length and thus in the wavelength of their absorption peak. It is therefore recognized that infrared sensitizing dyes are inclined to suffer from low and temperature-dependent efficiency of spectral sensitization, because of the low energy level of the LUMO in these molecules.

IV. INFRARED-SENSITIVE PHOTOGRAPHIC MATERIALS

Various kinds of infrared-sensitive photographic materials are now produced by several photographic companies for recording systems (a) with new light sources such as semiconductor lasers, which provide compact, stable, and cheap infrared beams for scanning, and (b) with detecting and discriminating capability which is enhanced by their sensitivity to infrared light.

In case (a), infrared-sensitive black and white films and papers are used as recording materials for scanners with semiconductor laser diodes in the fields of graphic arts and medical diagnosis. Figure 4 shows the spectral response of such materials (Fuji LD Scanner Film PD-100).

Figure 5 shows the spectral response of infrared-sensitive black and white materials (Kodak High Speed Infrared Film) for case (b).

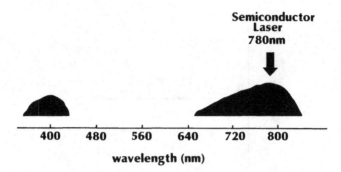

Figure 4. Wedge spectrum of Fuji LD Scanner Film, PD-100.

Figure 6 shows the composition and color-forming process of an infrared color negative film, in which there are three emulsion layers sensitive to green, red, and infrared light, respectively. A yellow filter is used to absorb blue light incident to the film, since all the emulsion layers are sensitive to blue light. Infrared, red, and green light incident to the film form cyan, magenta, and yellow dyes, respectively, in it after processing, and red, green, and blue images, respectively, in a color print. In a color film subjected to reversal processing, infrared, red, and green light form red, green, and blue images, respectively. A color given by an infrared color film is completely different from that in the original image and is called "false color." For example, Figure 7 shows the spectral response of an infrared-sensitive color film.

It is obvious that the introduction into photographic materials of sensitivity to infrared light, to which human beings are not sensitive, enhances their detecting and discriminating ability.

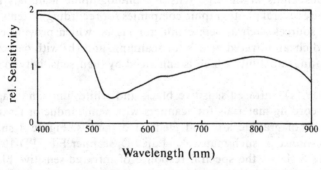

Figure 5. Sensitivity spectrum of Kodak High Speed Infrared Film.

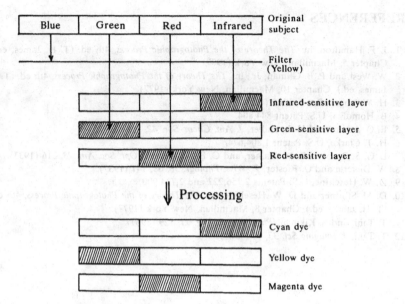

Figure 6. Composition of infrared color negative film.

V. FUTURE PROSPECTS

As stated above, infrared photography has a long history and is still important for various purposes. However, the stability and sensitizing ability of infrared sensitizing dyes used for it are far from ideal. A great deal of effort will be needed to develop infrared photography to successfully meet the demands.

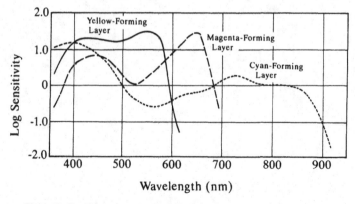

Figure 7. Sensitivity spectrum of Kodak Ektachrome Infrared Film.

REFERENCES

1. J. F. Hamilton, in: *The Theory of the Photographic Process*, 4th ed. (T. H. James, ed.), Chapter 5, Macmillan, New York (1977).
2. W. West and P. B. Gilman, Jr., in: *The Theory of the Photographic Process*, 4th ed. (T. H. James, ed.), Chapter 10, Macmillan, New York (1977).
3. H. W. Vogel, *Ber. 6*, 1302 (1873).
4. B. Homolka, U.S. Patent 844,804.
5. E. Q. Adams and H. L. Haller, *J. Am. Chem. Soc. 42*, 2661 (1920).
6. H. T. Clarke, U.S. Patent 1,804,674.
7. L. G. S. Brooker, F. M. Hamer, and C. E. K. Mees, *J. Opt. Soc. Am. 23*, 216 (1933).
8. W. Dieterle and O. Riester, *Z. Wiss. Photogr. 36*, 68, 141 (1937).
9. D. W. Heseltine, U.S. Patents 2,756,227 and 2,734,900.
10. D. M. Sturmer and D. W. Heseltine, in: *The Theory of the Photographic Process*, 4th ed., (T. H. James, ed.), Chapter 8, Macmillan, New York (1977).
11. T. Tani, and S. Kikuchi, *Photogr. Sci. Eng. 11*, 129 (1967).
12. T. Tani, *J. Imaging Sci. 33*, 17 (1989).

15

Medical Applications

ETHAN STERNBERG and DAVID DOLPHIN

I. HISTORY OF PHOTOSENSITIZERS

Plants and bacteria capture light in the visible and near-infrared region by absorption using various chlorophylls, while photosynthetic bacteria utilize bacteriochlorophylls. The resulting energy of the photoexcited state is ultimately utilized to execute a series of reductions of carbon dioxide and the liberation of oxygen from water.[1] Other photoexcited-state species, which are derived from compounds that absorb in the near-UV region, are utilized by plants as a means of protection against deprivation by fungi or bacteria or ingestion by animals or insects.[2,3] These near-UV and visible absorbing compounds act as photocatalysts for the conversion of triplet state oxygen to toxic, singlet state oxygen[4] (Figure 1, Type II) or may react in the absence of oxygen by photoinitiated electron transfer in the presence of a biological enzymatic system[5] (Figure 1, Type I), while the psoralens can cause DNA cross-linking.[6] A form of the latter has been approved as a phototherapy for a form of T-cell lymphoma.[7] This review will concentrate on Type I and Type II toxic modalities with compounds that absorb on the edge of the visible region and into the near-infrared region.

Investigation of photoeffects of dyes on animals and humans goes back to the studies of Raab in the latter part of the 19th century.[8] Policad in 1925 investigated the ability of porphyrins which included hematoporphyrin to initiate a phototoxic effect.[9] Exposure of this type in humans can result

ETHAN STERNBERG and DAVID DOLPHIN • Department of Chemistry, University of British Columbia, Vancouver, British Columbia, Canada V6T 1Y6.

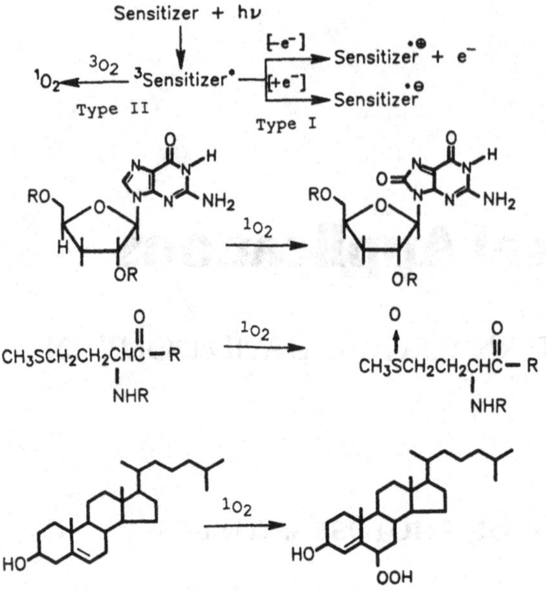

Figure 1. Type I and Type II reactions, and singlet oxygen reaction with guanosine, methionine, and cholesterol.

in severe skin necrosis that resembles porphyria. By 1942, it was recognized by Auler and Banzer that hematoporphyrin could be absorbed into cancerous (neoplastic) tissue.[10] Due to its fluorescent properties, the extent of hematoporphyrin incorporation could be qualitatively detected in neoplastic, damaged, and embryonic tissue.[11] In a most important discovery, Lipson *et al.* in 1964 recognized that an oligomerized derivative of hematoporphyrin (HpD) preferentially accumulated into cancerous tissue over surrounding tissue.[12-14] A purified version of this derivative, Photofrin II,* has been shown by Dougherty to possess certain clinical advantages over HpD for photodynamic therapy (PDT).[15]

Singlet oxygen has the ability to react with a wide variety of biochemical molecules.[16] These include guanosine, methionine, and cholesterol (Figure 1) and unsaturated compounds including lipids. The intra- or extracellular formation of singlet oxygen in sufficient quantities can cause significant damage and even death to the species unfortunate to ingest a sensitizer and then be exposed to light. The conversion of the relatively inert ground-state triplet oxygen to the very reactive singlet oxygen requires light energy in

---

* Photofrin II is marketed by Quadra Logic Technologies, Vancouver, British Columbia.

the range of 1270 nm (23 kcal mol⁻¹).[17] Many compounds that absorb in the near-UV and the visible region of the spectrum act as photosensitizers and accomplish the triplet–singlet interconversion of oxygen from their long-lived triplet excited states, as shown in Figure 1. This is known as a Type II photoprocess.[18] Until recently, little was known about this ability with regard to compounds which absorb into the near-IR region.

A review of the standard Yablonski diagram (Figure 2) shows the initial absorption of light excites the chromophore to the first excited-state singlet. Fluorescence can occur from this state, and this can be detected in tissue. The disallowed intersystem crossing may be catalyzed in the presence of trace metallic impurities. The resulting triplet of the sensitizer transfers energy by exciplex complex to the ground-state triplet of oxygen. This results in the production of singlet oxygen. From the diagram, one suspects that as long as the triplet lifetime of the sensitizer is long enough and energy of the state is greater than 1270 nm, singlet oxygen would be produced. In the absence of a reactive species, the singlet oxygen loses its energy through luminescence emission at 1270 nm. A semiconductor photodiode which detects at 1270 nm can measure this luminescence.[17]

Photoinitiated electron-transfer mechanisms are better known as Type I photoprocesses.[19] The carbon fixation cycle of plants is a highly sophisti-

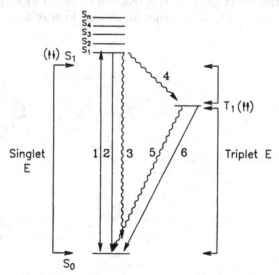

Figure 2. Yablonski diagram. 1, Absorption: depends on sensitizer; 2, fluorescence: lifetime depends on heavy metals and molecular interaction; 3, nonradiative loss of energy by molecular interaction; 4, nonradiative loss of energy by spin state interaction with metal impurities; 5, nonradiative loss of energy by molecular interaction; 6, phosphorescence.

cated version of this mechanistic protocol. The photoexcited state of the sensitizer can react catalytically in the presence of a highly oxidized or reduced compound by respectively delivering an electron to generate its radical cation or abstracting an electron to give a radical anion. This radical cation can go on to abstract an electron while this radical anion can go on to donate an electron. If these photoinitiated reactions occur in a sensitive area of the cell such as an enzyme in the mitochondria, then important metabolic processes will be disrupted.[20]

Photofrin II is the only singlet oxygen photosentizer approved for clinical studies in humans by the United States Food and Drug Administration. To date, over 4,000 patients have received Photofrin II in clinical trials.

II. HEMATOPORPHYRIN DERIVATIVE, PHOTOFRIN II, AND THE PRESENT

Any review into the field of near-infrared absorbing compounds and their medical applications toward the treatment of cancer and other diseases must include a summary of the history and medical applications of Photofrin II and hematoporphyrin derivative (HpD).[21] Both HpD and Photofrin II are synthesized by the addition of sulfuric acid to an acetic acid solution of hematoporphyrin (Figure 3). This reaction results in a mixture of mono-oligomeric acetates, which are neutralized with hydroxide. In the case of

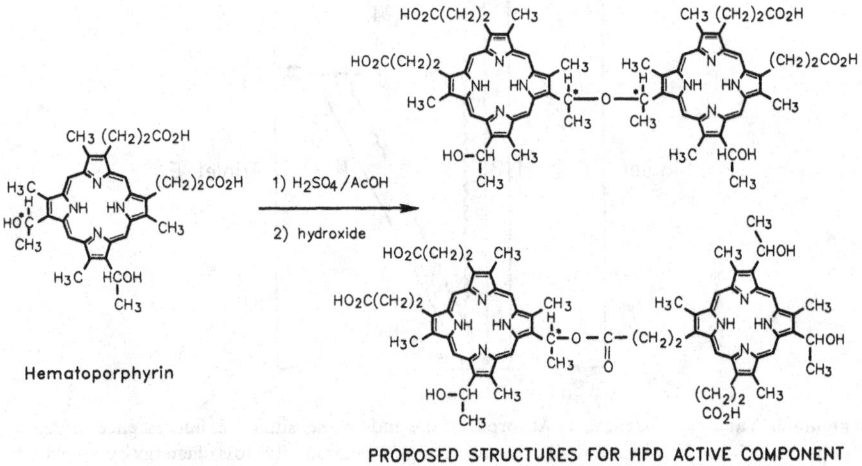

PROPOSED STRUCTURES FOR HPD ACTIVE COMPONENT

Figure 3. Ester- and ether-linked hematoporphyrin derivatives.

Photofrin II, the resulting mixture is purified so as to remove a majority of monomeric materials. The monomer seems to account for some of the skin photosensitivity, a clinical complication of PDT.[22]

It was suspected for many years that the active ingredient in this mixture was either an ester- or an ether-linked dimer.[23] Kessel *et al.* synthesized an ester-linked dimer and have found it to have some of the tumor accumulation properties of HpD (Figure 3).[24] Recent syntheses by Dougherty's group of dimeric and trimeric ether-linked materials have shown that the most active components for tumor treatment are not necessarily the dimers, but the trimers and probably other oligomers.[25]

The chemical structure is important in the biodistribution of this mixture into tumors. Just as important may be the nonbonding interactions between porphyrin molecules.[26] The resulting higher molecular weight aggregates in aqueous solution form unidentifiable components which may account for biodistribution and transport by low-density lipoproteins *in vivo*.[27] Binding sites for low-density lipoproteins are known to be high in cancerous tissue. In general, porphyrins, chlorins, and phthalocyanines all can be observed by uv–vis spectroscopy to aggregate extensively in aqueous solution (Figure 4).

Photofrin II has been utilized in the treatment of various cancers with varying degrees of success in over 4,000 patients. The material poses three problems with regard to therapy. The first is skin photosensitivity for up to

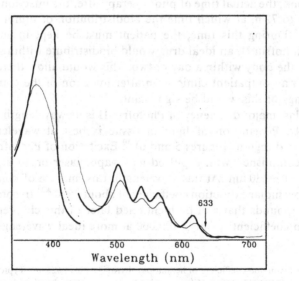

Figure 4. Comparison of visible spectra of (1) hematoporphyrin and (2) Photofrin II.

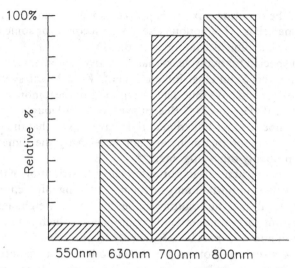

Figure 5. Relative penetration in muscle tissue of light at different wavelengths. For 800 nm, approximately 90% loss of intensity occurs for 8 mm of penetration.

two months after injection (~2 mg of compound per kilogram of body weight).[28] Several methods for reducing skin sensitivity after therapy by addition of singlet oxygen traps such as vitamin E have been proposed.[29]

Second, the actual time of phototherapy after the injection of Photofrin II ranges to 72 h, at which time the biodistribution of tumor to muscle is optimal.[28] During this time, the patient must be kept in subdued light. Unlike Photofrin II, an ideal drug would biodistribute within several hours and clear the body within a day or two. This would allow therapy of major cancers in an outpatient clinic soon after injection of the compound. The cost savings of this would be significant.

The last major deficiency of Photofrin II is its wavelength of excitation (Figure 4). Penetration of light in tissue is best at wavelengths in the near-infrared region (Figures 5 and 6).[30] Excitation of Photofrin II is most easily accomplished with a pulsed gold-vapor laser or an argon-pumped dye laser at ~630 nm. At this wavelength, this mixture of compounds has a rather low molar extinction coefficient of about 2500.*[31] In contrast, several other compounds that will be mentioned later in this chapter have molar extinction coefficients of near 100,000 at more ideal wavelengths for tissue

---

* The molar extinction coefficient of hematoporphyrin is 4350 at 623 nm. Typically, aggregated porphyrins such as Photofrin II have peaks that are red-shifted with reduced intensity.

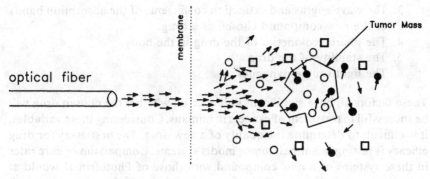

optical fiber

O — chromophore (specific for laser)

↑ — light (specific wavelength)

□ — chromophore (general)

● — activated chromophore

Figure 6. Optical fibre irradiating a tumor. Isotropic scattering is shown.

penetration. Although an argon-pumped laser could also be used for these chromophores, it would be preferable to use the new high-power, low-cost diode lasers which radiate in the region between 780 and 810 nm. Also, these lasers could be used to pump a modified YAG laser to generate light in the 660-nm region.

One of the most attractive aspects of this therapy is the process by which Photofrin II is visualized during treatment by fluorescence.[32-34] The level of fluorescence at about 670 nm in the tumor as compared to that in background tissue may reach a factor of greater than 10. This is remarkable in that the overall biodistribution of the compound as determined by radioactive labeling is about 3:1.[35] The actual distribution of the Photofrin II in the tumor may account for this difference. A larger portion of the compound seems to locate in the vasculature and at the surface of the tumor,[38] where it is easily detected. Greater photodamage to the vasculature of a tumor would be expected under these conditions, thus severely limiting blood supply. This may mean that total cell irradiation is not necessary to eliminate the tumor.

The success of a new drug in the field of photodynamic therapy will be dependent upon the following:

1. The efficiency with which a photosensitizer converts triplet oxygen to singlet oxygen.
2. The optical properties of human tissue at various wavelengths.

3. The wavelengths and extinction coefficients of the absorption bands of the new compound chosen as a drug.
4. The pharmacokinetics of the drug in the body.
5. The stability of the drug.
6. The light source dynamics.

These factors all have significant influence on whether the chosen drug will be successful in treatment of cancers in humans. Considering these variables, it is difficult to determine the quality of a new drug. The best assay for drug efficacy is testing in animal/tumor model systems. Comparison of cure rates in these systems of a new compound with those of Photofrin II would at least give some indication of the future usefulness of the new compound. This type of comparison will not always be the case in this review.

Photodynamic therapy utilizing Photofrin II has been applied to a remarkable array of cancers. These include head and neck cancer, cutaneous and subcutaneous malignancies, cancer of the bladder, endobronchial tumors, cancers of the central nervous system, some eye tumors, gynecological malignancies, bronchial/lung tumors, and esophageal and gastrointestinal malignancies.[37] Many of the patients who have been treated with Photofrin II have been treated in the advanced stages of the disease. Without question, in these cases palliative treatments have been achieved. Only recently have some patients been treated early in the disease. Among these, many regressions and possible cures have been achieved. This is remarkable in that many of these cancers, such as those of the lung and the esophagus, have very low rates ($\sim$1%) of survival.

III. NEW COMPOUNDS

So far, this chapter has been concerned with giving an overview of the history and effectiveness of photodynamic therapy. New compounds for PDT which absorb on the edge of the near-infrared region and into it are being investigated by a number of groups throughout the world. One statement can be made about these "new sensitizers." Unlike Photofrin II, they all have definite structural integrity, thus making both *in vitro* and *in vivo* studies somewhat more facile. We have chosen to divide these compounds into three categories:

1. The cationic dyes.
2. The phthalocyanines.
3. The modified natural porphyrins, chlorins, and bacteriochlorins.

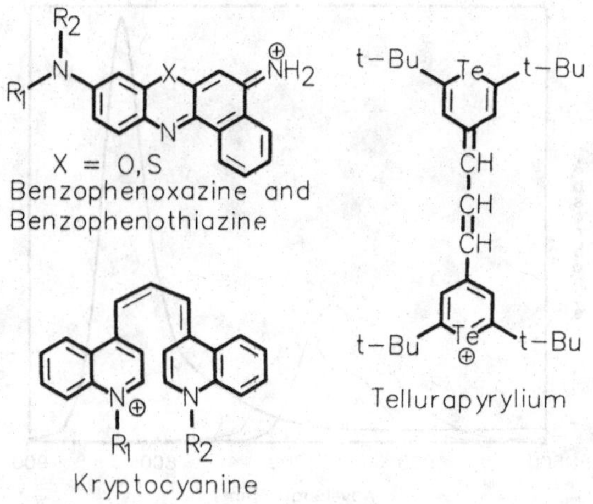

Figure 7. Structures of benzo[a]phenoxazinium and benzo[a]phenothiazinium dyes, kryptocyanine dyes ($R_1 = R_2$ = 2-ethyl-1,3-dioxolane, EDCK), and a chalcogenapyrylium (tellurapyrylium) dye.

A. Charged Dyes

Cationic dyes are just beginning to be extensively investigated as Type I and Type II phototoxins. These compounds can be modified to absorb light in the near-infrared region.

Examples of these include benzo[a]phenoxazinium and benzo[a]-phenothiazinium cationic dyes (Figure 7).[38] Cincotta and his co-workers have shown that the absorption spectra are dependent upon the substituents and the state of protonation of the dye. The visible absorption spectra generally range from 620 to 660 nm with molar extinction coefficients of ~80,000 (Figure 8). The increased efficiency of these compounds for killing cells over that of Photofrin II had not been established in these studies. However, several compounds were evaluated against a number of cell lines including HeLa-carcinoma cells, Hep 2-larynx carcinoma, and the normal cell line of BFK-bovine kidney. After various times of incubation with the compounds, the cells were irradiated for 10 min with a tungsten lamp source giving power of 8 mW cm^{-2} at 33 cm. Efficiency of killing was dependent upon the pK_a of the dye. A low pK_a of ~5.0 resulted in poor photosensitivity. Surprisingly, dyes which generated high yields of singlet oxygen were not necessarily more effective at cell killing.

If these dyes follow the general trend for cationic dyes, mitochondrial toxicity may be responsible for cell death.[39,40] Mitochondrial staining and eventual disruption of the enzymatic systems may not be singlet oxygen

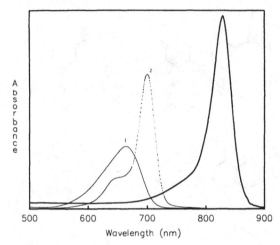

Figure 8. Visible–near-infrared spectra of (1) benzo[a]phenoxazinium, (2) kryptocyanine, and (3) tellurapyrylium dye.

related. It has been found that in the absence of oxygen the cationic dyes are still phototoxic.[41] More investigation into the metabolic processes that are disrupted by these types of phototoxins is needed.

Of the cationic dyes, the kryptocyanine dyes have been evaluated to the greatest extent with regard to their use in PDT[42,43] (Figure 7). These compounds have found use in dye lasers and can be modified by extending conjugation to absorb well into the near-infrared range* (Figure 8). Practically, these polyene-modified dyes are too unstable toward singlet oxygen to be of great use.

In vitro studies of N,N'-bis(2-ethyl-1,3-dioxolane)kryptocyanine (ECK) (Figure 7) showed that the main source of phototoxicity resides in the mitochondria.[43,44] Toxicity seemed to be related to a Type 1 mechanism because no dependence was found that was related to singlet oxygen lifetime or generation. It was observed in the studies that this effective class of phototoxins requires a great deal of light to elicit cell death. It was suggested that this fact was due to the low intracellular concentrations of the compounds as determined by fluorescence. Further *in vitro* studies found that the rate of photobleaching of the compounds was not singlet oxygen dependent.

In vivo studies with EDCK indicated no cures.[44] Quite high concentrations of the dye (~10 mg/kg) showed very good preferential accumulation

* Laser dyes available from Kodak, Rochester, New York.

in the tumor over skin and muscle. Upon photoirradiation with very intense light, low or no skin toxicity was observed, while tumor cell death reached 80–90%. However, the actual concentration of the dye in the cell, specifically the mitochondria, was rather low compared to that of compounds like Photofrin II. Oseroff *et al.* suggested a regime that would include both dye and Photofrin II or chlorin e_6/monoclonal antibodies as a more effective treatment protocol.[43,44]

The chalcogenapyrylium dyes first developed by Detty and Loss at Eastman Kodak are structurally similar to the kryptocyanine dyes.[45,46] The major change in the behavior of these chromophores results from the substitution of the charged nitrogen in a pyridinium ring with either oxygen, sulfur, selenium, or tellurium (Figure 7). The resulting bathochromic shifts can range up to 50 nm (Figure 8). This property, along with changing the substituents on the ring to protect it from hydrolysis, has been utilized in designing relatively stable dyes that absorb at the ideal wavelengths (780–820 nm, molar extinction coefficients of up to 300,000) for excitation by the diode lasers. Work so far suggests that these dyes readily pass the blood-brain barrier to effectively concentrate in tumors of the central nervous system.[47]

Biological studies by Powers indicated that toxicity by the chalcogenapyrylium dyes was probably not the result of singlet oxygen, as determined by flushing the *in vitro* assay with nitrogen.[48] After exposure of these cells to a 1.0-μM solution of the dye, toxicity occurred in the dark preferentially to tumor cells over normal cells by a factor of 2–3. Toxicity was increased by exposure to light. A linear relationship between the instability of the dye and the toxicity to various cell lines was noted. Powers has stated that he is presently looking for analogues with increased chemical stability along with phototoxicity.[49]

B. Phthalocyanines

Phthalocyanines were first accidentally synthesized by combining phthalic anhydride and ammonia at Scottish Dyes Limited in 1928.[50] In this case, the iron phthalocyanines were isolated. Notable properties of phthalocyanines are their intense color and incredible stability to light, acid, and oxygen.[51] Another notable feature of phthalocyanines and metallophthalocyanines is their very low solubility in all solvents. Generally, most of the dyes derived from the phthanocyanines are aggregates. Substituents on the aromatic ring such as SO_3^- and CO_2^- produce water-soluble complexes. Metalated, sulfonated phthalocyanines have been studied most extensively for use in PDT (Figure 9). Various metal complexes result in different levels of effectiveness. Other derivatives that have been investigated for PDT of late are the metalated naphthalocyanines.

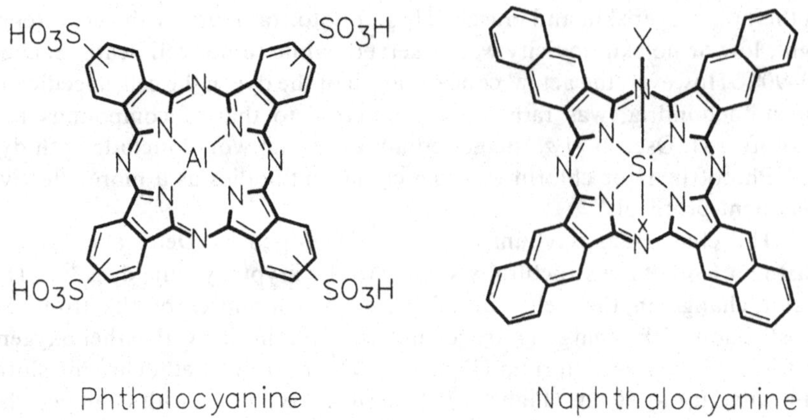

Phthalocyanine   Naphthalocyanine

Figure 9. Structures of aluminum tetrasulfonated phthalocyanine and silicon naphthalocyanine.

Because of the interest in phthalocyanines as catalysts and photocatalysts, a large amount of literature exists about their photochemical triplet lifetimes. For the metalated naphthalocyanines (Zn, Si, Al), where four aromatic rings have been fused onto the parent molecule, these range from 100 to 300 μs while the absorption spectra range from 769 to 776 nm (molar extinction coefficients of ~100,000)[52] (Figure 10). It is interesting to note that the luminescence spectra of some of these compounds range

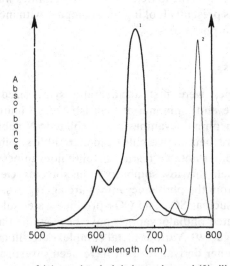

Figure 10. Visible spectra of (1) metalated phthalocyanine and (2) silicon naphthalocyanine.

beyond 1270 nm. Because of this, one suspects that the conversion of triplet oxygen to singlet oxygen may occur through a thermally assisted exiplex reaction.[53] Early *in vitro* studies on silicon dihexylalkoxynaphthalocyanine have shown it to be very effective at photolytic cell killing.[54] Very early *in vivo* studies with these compounds utilizing a diode laser as a light source have found good efficacy.[55]

Recent years have seen a large increase in the number of papers being published on the possible use of phthalocyanines in PDT.[56] A number of metalated, sulfonated, hydroxylated, and alkoxylated derivatives have shown good celling abilities *in vitro*.[57-59] These compounds absorb in the region of 670–690 nm with molar extinction coefficients of near 10^5 and triplet lifetimes ranging up to 500 μs (Figure 10). Most of the *in vitro* studies have shown that the Zn and Al complexes are most effective at eliciting cell death when exposed to light. Further studies show that the source of this toxicity is singlet oxygen although it is known that phthalocyanines can catalyze Type 1 electron-transfer reactions.[60] Presently, the compounds which have been studied to the greatest extent are the sulfonated zinc complexes. Work by van Lier and co-workers has indicated that the most useful compound may be the disulfonated derivative.[61,62]

Biodistribution studies by Chan *et al.* utilizing fluorescence of the disaggregated tissue showed that the chloroaluminum sulfonated phthalocyanines were taken up preferentially by tumors over normal tissues such as muscle and skin.[63] In the same paper, the weights of four types of malignant tumors treated *in vivo* with the above phthalocyanine and exposed to light were significantly reduced upon removal at five days as compared to a control. In a study by van Lier and co-workers, it was observed that this compound along with the gallium and cerium complexes can also initiate cures on the same level as Photofrin II.[64]

No large inherent toxicity has been found for most of the phthalocyanines or the naphthalocyanines. However, some toxicity of cell lines maintained in the dark has been indicated *in vitro* for the aluminum nonsulfonated and sulfonated phthalocyanines at rather high dosage ranges.[64] These lie well beyond those used for therapy. Because of their ease of synthesis and their general effectiveness, the future looks good for these compounds as agents for PDT.

C. Chlorins

Simple derivatives of tetrapyrrole compounds derived from animal sources such as hematoporphyrin and protoporphyrin have been extensively investigated in the field of photodynamic therapy. Derivatives of the natural plant pigments, such as the chlorins and bacteriochlorins (Figure 11), have

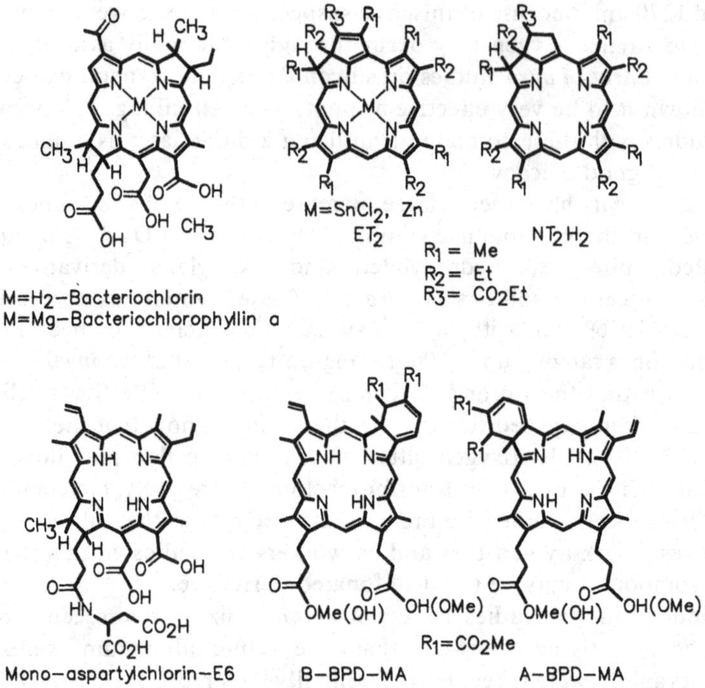

Figure 11. Structures of chlorin e_6, bacteriochlorophyllin and bacteriochlorin a, the purpurins ET_2 and NT_2H_2, and A- and B-ring benzoporphyrin derivatives.

only recently been investigated. This seems to be a most promising area of research for the development of new compounds for PDT. Like derivatives of hematoporphyrin (Photofrin II), these compounds seem to have either little or no inherent toxicity in dose ranges well beyond those necessary to achieve 100% photolytic cell death *in vitro*. To our knowledge, no nonphototoxic effects have been noted *in vivo* with any of the plantlike derivatives mentioned below. While large-scale synthesis of dyes or phthalocyanines is relatively easy, isolation of large quantities of the natural phototoxins or synthesis of analogues of the chlorins or bacteriochlorins is not so straightforward. Synthetic derivatives hold the most promise for generating the quantities necessary for clinical trials.

The natural pheophorbide *a* (670 nm; structure and spectra not shown)[65-68] and bacteriochlorin *a* (770 nm) were investigated for phototherapy (Figure 12).[69,70] *In vitro* studies showed that pheophorbide *a* has approximately the same phototoxicity as HpD, while bacteriochlorin *a* seemed to be more effective than HpD at killing cells. *In vivo* work with

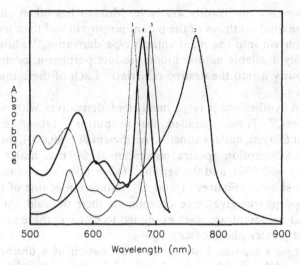

Figure 12. Visible–near-infrared spectra of (1) chlorin e_6, (2) bacteriochlorin a, (3) NT$_2$H$_2$, and (4) benzoporphyrin derivative.

pheophorbide a indicated that it is effective at causing tumor shrinkage in an N-methyl-N-nitrosourea-induced rat tumor system.[70]

The natural chromophore chlorin e_6 and its more soluble synthetic mono-aspartic acid derivative absorb at ~664 nm (molar extinction coefficients of ~25,000) (Figures 11 and 12).*[73] Berns and co-workers observed good uptake into tumors, especially with the monoaspartyl derivative of chlorin e_6.[71] Low skin photosensitivity was also apparent from studies of these compounds.[72-74] The results of Berns and co-workers indicated that a 100% cure rate can be achieved with dosages of 8 mg of compound per kilogram of weight in mice with an EMT-6 undifferentiated sarcoma. In a series of 12 animals, two were sacrificed immediately after irradiation to characterize the level of necrosis in the tumor.[71] Dosages of less than 8 mg per kilogram showed less significant damage in these assays. In studies of skin sensitivity of animals treated with 10.0 mg of the monoaspartyl derivative per kilogram with a light dosage of 10 to a high 100 J cm^{-2}, none of the typical skin necrosis associated with HpD was seen.

Synthetic derivatives of chlorin type molecules have only recently been investigated in the field of PDT. Work by Alan Morgan at Toledo and our group at the University of British Columbia and Quadra Logic Technologies was inspired by the fact that large quantities of the natural plant

* A source of derived chlorin-e_6 is Bruce Burnham, Porphyrin Products, Inc., Logan, Utah.

chromophores are not readily available. Morgan has taken an approach involving the total synthesis of the parent porphyrin and then modification of that porphyrin into the plant chlorin type derivative,[74] while we begin with a readily available natural blood product porphyrin, protoporphyrin IX, and modify it into the desired material.[75] Each of these routes has its advantages.

Morgan synthesized a large number of derivatives with a variety of chromophores.[74] These included the purpurins[76] (absorption spectra maximum at 690 nm, molar extinction coefficient of ~40,000), the metalated purpurins[77] (absorption spectra maximum at 690 nm, molar extinction coefficient of ~80,000), and the verdins[78] (absorption spectra maximum of 695 nm, not shown) (Figures 11 and 12). Saturation of one of the pyrrole rings of a porphyrin give these compounds their unusual chromophore. Both Zn and Sn complexes were examined to discover these modifications would effect greater phototoxicity.

In Morgan's studies, histological examination of a urothelial tumor model induced by N-[4-(5-nitro-2-furfuryl)]-2-thiazoyl which had been grafted subcutaneously into rats was performed 4 and 24 h after treatment with a number of purpurins and metalated purpurins (10 mg/kg).[79,80] Based on these initial results, six compounds were chosen for further dose studies. Also, tumor blood flows were examined for $SnET_2$ (5 mg/kg) by tracing the progress of radioactive spheres. This study showed that tumor blood flow was significantly reduced after photoradiative therapy.

Studies in which doses ranged from 0.5 to 2.5 mg showed that the various metalated purpurins $ZnET_2$, $SnET_2$, and $SnNT_2H_2$ (Figure 11) caused tumor regression for 12 days. Further studies showed that after photoirradiation (360 J cm^{-2}) 70% of animals were tumor-free 30 days after treatment with $SnET_2$. Also, it was indicated that the most effective time of phototreatment was 24 h after injection. This minimized skin edema while maintaining sufficient levels of compound in the tumor for treatment. A most interesting result was obtained with the R3327 cell line in rats. It was known to be resistant to gamma-radiation treatment. It was also found to be resistant to treatment by similar levels of the phototoxins used successfully above.

The benzoporphyrin derivatives (BPDs; see Figure 11) are easily obtained by a Diels–Alder reaction between the readily available diene protophorphyrin IX dimethyl ester and the strong dienophile dimethylacetylene dicarboxylate.[75] Separation of the resulting geometric isomers (A- and B-ring adducts) is followed by isomerization of the 1,4-diene system to the 1,3-diene system. Dependent upon solvent, the molar extinction coefficient reaches 33,000 at 690 nm for these compounds (Figure 12). Partial or total hydrolysis of the propionic esters results in, respectively, mono- or diacid derivatives.

In vitro studies found that the monoacid derivatives were more cytotoxic by a factor of 40 on an M1 cell line than Photofrin II.[81] *In vivo* biodistribution studies were initiated over a series of times ranging from 3 to 168 h for the tritium-labeled BPD monoacid A-ring isomer.[82] They showed that the compound preferentially binds to tumor over normal muscle tissue and skin by factors of 2.68 and 2.57, respectively, for mice seeded with p815 tumors. The results for muscle are comparable to Photofrin II biodistributions while the ratio for skin is much higher.

In vivo biodistribution studies of the tritium-labeled BPD monoacid A-ring isomer for times ranging from 3 to 16 h found that the best tumor/skin/muscle ratio was already achieved at 3 h. It was determined from these biodistribution studies and a series of phototreatments with light (378 J cm^{-2}) at 3 and 24 h that earlier treatment is more appropriate with the BPD compounds.[83,84] Overall, the above indicated that treatment would be possible shortly after injection (within 3 h) for this photosensitizer, and resultant long-term photosensitivity would be substantially less. Efficacy studies have shown that the BPDs are at least as effective as Photofrin II at initiating cures in mice in the only assay that is approved by the FDA, while some of the isomeric compounds studied are possibly better.

IV. FUTURE TRENDS IN PHOTODYNAMIC THERAPY

We have shown in this review that photosensitizers derived from plants, animals, and synthetic sources have formed the basis for a new field of therapy for the treatment of cancer. Each of the three types of sensitizers reviewed herein has unique properties that have yet to be explored. All mimic the known photochemical behavior of plants or photosynthetic bacteria.

The future direction of research in the field of photodynamic therapy spans several disciplines and areas. Already, indications are that modifications of this technology can be used for removing viruses and parasites from blood,[85] help cure psoriasis and other skin diseases,[86] utilize monoclonal antibodies as transport agents for the phototoxin,[87] take advantage of the compound's inherent fluorescence for detection of very small areas of tumorous cells, and finally be applied to the clearing of cardiac plaques.[88] Only one's imagination limits the areas of medical application in which PDT can be applied.

ACKNOWLEDGMENTS

We wish to acknowledge the assistance of Chand Sishta, David Thackray, and Jack Chow in the preparation of this manuscript.

REFERENCES

1. R. K. Clayton, *Light and Living Matter: A Guide to the Study of Photobiology*, McGraw-Hill, New York (1970).
2. J. B. Hudson and G. H. N. Towers, *Photochem. Photobiol. 48*, 289 (1988).
3. N. Duran and P. S. Song, *Photochem. Photobiol. 43*, 677 (1986).
4. I. Saito and T. Matsiura, eds., *Tetrahedron 41*, 2037 (1985).
5. N. J. Turro, *Molecular Photochemistry*, W. A. Benjamin, New York (1965).
6. H. Bayley, F. Guspurro, and R. Edelson, *TIPS 8*, 138 (1987).
7. M. F. Edelson, *Sci. Am.*, 68 (August 1988).
8. O. Raab, *Z. Biol. 19*, 524 (1900).
9. A. Policad, *C. R. Hebd. Soc. Biol. 91*, 1422 (1924).
10. H. Auler and G. Banzer, *Z. Krebsforsch. 53*, 65 (1942).
11. F. H. J. Frigge, G. S. Wieland, and L. O. J. Mungariello, *Proc. Soc. Exp. Biol. Med. 68*, 640 (1948).
12. R. L. Lipson, E. J. Blader, and E. M. Olsen, *J. Natl. Cancer Inst. 26*, 1 (1961).
13. R. L. Lipson, M. J. Gray, and E. J. Blades, Proceedings of the IX International Cancer Conference, p. 393 (1966).
14. T. J. Dougherty, *Cancer Res. 38*, 2628 (1988).
15. T. J. Dougherty, *Photochem. Photobiol. 46*, 569 (1987).
16. H. Seis, *Angew. Chem. Int. Ed. Engl. 25*, 1058 (1986).
17. C. S. Foote, in: *Progress in Clinical and Biological Research*, Vol. 170, *Porphyrin Localization and Treatment of Tumors* (D. R. Doiron and C. J. Gomer, eds.), p. 3, Alan R Liss, New York (1984).
18. D. R. Doiron, in: *Progress in Clinical and Biological Research*, Vol. 170, *Porphyrin Localization and Treatment of Tumors* (D. R. Doiron and C. J. Gomer, eds.), p. 46, Alan R. Liss, New York (1984).
19. E. I. Kapinus, M. M. Weksankia, V. P. Starzi, V. I. Boghillo, and I. I. Dilung, *J. Chem. Soc., Faraday Trans. 2, 81*, 631 (1985).
20. S. V. Powers, *Proc. SPIE–Int. Soc. Opt. Eng. 847*, 77 (1987).
21. D. Kessel, ed., *Advances in Experimental Medicine and Biology*, Vol. 193, *Methods in Porphyrin Photosensitization*, Plenum Press, New York (1985).
22. D. Kessel and M. Chang, *Photochem. Photobiol. 47*, 277 (1988).
23. T. J. Dougherty, *Photochem. Photobiol. 46*, 569 (1987).
24. D. Kessel, B. Musselman, and C. K. Chang, *Photochem. Photobiol. 46*, 563 (1987).
25. R. Pandey, B. Bellinger, K. Ho, and T. Dougherty, *Lasers in Medical Science 3*, Abstract 52 (1988).
26. D. Dolphin and T. P. Wijesekara, in: *Advances in Experimental Medicine and Biology*, Vol. 193, *Methods in Porphyrin Photosensitization* (D. Kessel, ed.), p. 241, Plenum Press, New York (1985).
27. D. Kessel, C. K. Chang, and B. Musselman, in: *Advances in Experimental Medicine and Biology*, Vol. 193, *Methods in Porphyrin Photosensitization* (D. Kessel, ed.), p. 213 Plenum Press, New York (1985).
28. T. J. Dougherty, in: *Adjuncts to Cancer Therapy* (S. G. Economou, ed.), Lea and Febiger, Philadelphia (1989).
29. T. A. Dahl, R. Miden, and P. E. Hartman, *Photochem. Photobiol. 47*, 357 (1988).
30. B. C. Wilson, W. P. James, D. M. Lowe, and G. Adams, in: *Progress in Clinical and Biological Research*, Vol. 170, *Porphyrin Localization and Treatment of Tumors* (D. R. Doiron and C. J. Gomer, eds.), p. 115, Alan R. Liss, New York (1984).
31. R. K. DiNello and C. K. Chang, in: *The Porphyrins, Part A* (D. Dolphin, ed.), Vol. 1, p. 290 (1978).

32. M. Zandomenghi, C. Fiesta, C. A. Angeletti, G. Menconi, T. Sicuro, and I. Cozzani, *Lasers in Medical Science 3*, Abstract 99 (1988).

33. O. J. Balchum, K. B. Patel, Y. N. Qun, and D. R. Doiron, in: *Progress in Clinical and Biological Research*, Vol. 170, *Porphyrin Localization and Treatment of Tumors* (D. R. Doiron and C. J. Gomer, eds.), p. 521, Alan R. Liss, New York (1984).

34. O. J. Balchum. A. E. Profio, D. R. Doiron, and G. C. Huth, in: *Progress in Clinical and Biological Research*, Vol. 170, *Porphyrin Localization and Treatment of Tumors* (D. R. Doiron and C. J. Gomer, eds.), p. 847, Alan R. Liss, New York (1984).

35. C. J. Gomer and F. M. Little, in: *Progress in Clinical and Biological Research*, Vol. 170, *Porphyrin Localization and Treatment of Tumors* (D. R. Doiron and C. J. Gomer, eds.), p. 591, Alan R. Liss, New York (1984).

36. B. M. Henderson, T. J. Dougherty, and P. B. Malone, in: *Progress in Clinical and Biological Research*, Vol. 170, *Porphyrin Localization and Treatment of Tumors* (D. R. Doiron and C. J. Gomer, eds.), p. 601, Alan R. Liss, New York (1984).

37. M. J. Manyak, A. Russo, P. D. Smith, and E. Glatstein, *J. Clin. Oncol.* 6, 380 (1988).

38. L. Cincotta, J. W. Goley, and A. Concotta, *Photochem. Photobiol.* 46, 751 (1987).

39. Z. Darzynkiewicz and S. Carter, *Cancer Res.* 48, 1295 (1988).

40. R. Hilf, S. L. Gibson, R. S. Murant, T. L. Ceckler, and R. G. Bryant, *Proc. SPIE – Int. Soc. Opt. Eng.* 847, 2 (1987).

41. P. Valdes-Aguileri, L. Cincatta, J. Foley, and I. E. Kochevar, *Photochem. Photobiol.* 45, 337 (1987).

42. A. R. Oseroff, D. Ohuoha, G. Ara, D. McAulifte, J. Foley, and L. Cincotta, *Proc. Natl. Acad. Sci. U.S.A.* 83, 9729 (1986).

43. A. R. Oseroff, G. Ara, D. Ohuoha, J. Aprille, J. C. Bommer, M. L. Yarmush, and L. Cincotta, *Photochem. Photobiol.* 46, 83 (1987).

44. A. Ara, J. R. Aprille, C. D. Malis, S. B. Kane, L. Cincotta, J. Foley, J. V. Bonventre, and A. R. Oseroff, *Proc. Natl. Acad. Sci. U.S.A.* 83, 8744 (1987).

45. M. R. Detty, *Proc. SPIE – Int. Soc. Opt. Eng.* 847, 68 (1987).

46. M. R. Detty and H. R. Loss, *Organometallics* 5, 2250 (1986).

47. S. K. Powers, D. L. Walstead, J. T. Brown, M. Detty, and P. J. Watkins, *J. Neuro Oncol.* submitted (1988).

48. S. K. Powers, *Proc. SPIE – Int. Soc. Opt. Eng.* 847, 74 (1987).

49. S. K. Powers, private communications (September 1988).

50. R. P. Linstead, *J. C. S.*, 1016 (1934).

51. F. H. Moser and A. L. Thomas, *The Phthalocyanines*, CRC Press, Boca Raton, Florida (1983).

52. I. McCubbins and D. Phillips, *J. Photochem.* 34, 639 (1986).

53. P. A. Firey and M. A. J. Rodgers, *Photochem. Photobiol.* 45, 535 (1987).

54. D. Wohrle, N. Isakandar, and A. Wendt, *Lasers in Medical Science 3*, Abstract 60 (1988).

55. M. A. J. Rodgers, private communication (October 1988).

56. P. Riesz and C. M. Krishna, *Proc. SPIE – Int. Soc. Opt. Eng.* 847, 15 (1987).

57. I. Rosenthal, E. Ben-Hur, S. Greenberg, A. Concepcion-Lam, D. M. Drew, and C. C. Leznoff, *Photochem. Photobiol.* 46, 959 (1987).

58. W. S. Chan, J. F. Marshall, R. Svensen, D. Phillips, and I. R. Hart, *Photochem. Photobiol.* 45, 757 (1987).

59. J. D. Spikes, *Photochem. Photobiol.* 43, 691 (1986).

60. T. Ohno, S. Kato, A. Yamada, and T. Tanno, *J. Phys. Chem.* 87, 775 (1983).

61. H. Ali, R. Langlois, J. R. Wagner, N. Bresseur, B. Paquette, and J. E. van Lier, *Photochem. Photobiol.* 47, 713 (1988).

62. B. Bresseur, H. Ali, R. Langlois, and J. E. van Lier, *Photochem. Photobiol.* 47, 705 (1988).

63. W. S. Chan, J. F. Marshall, G. Y. F. Lam, and I. R. Hart, *Cancer Res.* 48, 3040 (1988).

64. N. Bresseur, H. Ali, R. Langlois, J. R. Wagner, J. Rosseu, and J. E. van Lier, *Photochem. Photobiol. 45*, 581 (1987).
65. K. Iwai, M. Horigonie, and S. Kimura, *Photomed. Photobiol. 8*, 25 (1986).
66. T. Numata, S. Sukakibau, and N. Mizuno, *Photomed. Photobiol. 8*, 49 (1986).
67. N. Maeda, K. Ichikawa, and N. Mizuno, *Photomed. Photobiol. 8*, 183 (1986).
68. B. Roder, *Stud. Biophys. 114*, 183 (1986).
69. K. Irie, M. Shirosuki, T. Ando, S. Nakajima, N. Samejima, Y. Kakiuchi, I. Sakata, and K. Kashimizu, *Lasers in Medical Science 3*, Abstract 61 (1988).
70. E. M. Beems, T. M. A. R. Dubbelman, J. Lugtenburg, J. A. Van Best, M. M. F. A. Smeets, and J. P. T. Boegheim, *Photochem. Photobiol. 46*, 639 (1987).
71. J. S. Nelson, W. G. Roberts, and M. W. Berns, *Cancer Res. 47*, 4681 (1987).
72. K. Aizawa, T. Okuraka, T. Ohtani, H. Kawabe, Y. Yasunaka, S. O'Hata, N. Ohtoma, K. Bishimiya, C. Konaka, H. Kato, Y. Hagata, and T. Saito, *Photochem. Photobiol. 46*, 789 (1987).
73. W. L. Nourse, R. M. Parkhurst, W. A. Skinner, and R. T. Jordan, *Biochem. Biophys. Res. Commun. 151*, 506 (1988).
74. A. R. Morgan, A. Ramysersaud, G. M. Garbo, R. W. Keck, and S. H. Selman, *J. Med. Chem.*, submitted (1988).
75. P. Scherrer, A. R. Morgan, and D. Dolphin, *J. Org. Chem. 51*, 1094 (1986).
76. A. R. Morgan, S. Nonis, and A. Rampersand, *Proc. SPIE – Int. Soc. Opt. Eng. 847*, 166 (1987).
77. A. R. Morgan, G. M. Garbo, R. W. Keck, and S. H. Selman, *Proc. SPIE-Int. Soc. Opt. Eng. 847*, 172 (1987).
78. A. R. Morgan, A. Rampersand, R. W. Keck and S. H. Selman, *Photochem. Photobiol. 46*, 441 (1987).
79. A. R. Morgan, M. Kreimer-Birnbaum, G. M. Garbo, R. Q. Keck, and S. H. Selman, *Proc. SPIE-Int. Soc. Opt. Eng. 847*, 29 (1987).
80. A. R. Morgan, G. M. Garbo, R. W. Keck, and S. H. Selman, *Proc. SPIE-Int. Soc. Opt. Eng. 847*, 180 (1987).
81. A. M. Richter, B. Kelly, J. Chow, D. J. Liu, G. H. N. Towers, D. Dolphin, and J. G. Levy, *J. Natl. Cancer Inst. 79*, 1324 (1987).
82. A. M. Richter, S. Cerruti-Sola, E. D. Sternberg, D. Dolphin, and J. G. Levy, *J. Photobiol. Photochem.*, in press (1990).
83. A. M. Richter, E. M. Waterfield, A. K. Jain, E. D. Sternberg, D. Dolphin, and J. G. Levy, *J. Photobiol. Photochem.*, in press (1990).
84. A. M. Richter, E. Waterfield, E. Sternberg, D. Dolphin, and J. G. Levy, *Lasers in Medical Science 3*, Abstract 58 (1988).
85. J. L. Matthews, J. T. Newman, F. Sogundares-Bernal, M. M. Judy, H. Skiles, J. E. Levison, A. J. Marengo-Rowe, and T. C. Chan, *Transfusions 28*, 81 (1988).
86. J. L. McCullough, G. D. Wanstein, J. L. Douglas, and M. W. Berns, *Photochem. Photobiol. 46*, 77 (1987).
87. C. K. Wat, D. Mew, J. G. Levy, and G. H. Towers, in: *Progress in Clinical and Biological Research*, Vol. 170, *Porphyrin Localization and Treatment of Tumors* (D. R. Doiron and C. J. Gomer, eds.), p. 351, Alan R. Liss, New York (1984).
88. M. Vincent and W. Mackey, *Lasers in Surgery and Medicine 8*, Abstract 2 (1988).

Index